OXFORD MEDICAL PUBLICATIONS

Studies in economic appraisal in health care
Volume 2

Studies in economic appraisal in health care Volume 2

M. F. Drummond
*Senior Lecturer in Economics,
Health Services Management Centre,
University of Birmingham*

Anne Ludbrook
*Deputy Director,
Health Economics Research Unit,
University of Aberdeen*

Karin Lowson
*Treasurers' Department,
North Western Regional Health Authority*

Ann Steele
*Treasurers' Department,
Salford Health Authority*

Oxford New York Tokyo

OXFORD UNIVERSITY PRESS

1986

· Oxford University Press, Walton Street, Oxford OX2 6DP

Oxford   New York   Toronto
Delhi   Bombay   Calcutta   Madras   Karachi
Kuala Lumpur   Singapore   Hong Kong   Tokyo
Nairobi   Dar es Salaam   Cape Town
Melbourne   Auckland

and associated companies in
Beirut   Berlin   Ibadan   Nicosia

Oxford is a trade mark of Oxford University Press

Published in the United States
by Oxford University Press, New York

© Institute of Social and Economic Research, University of York, 1986

British Library Cataloguing in Publication Data
Studies in economic appraisal in health care.—
(Oxford medical publications)
Vol. 2
1. Medical care—Great Britain—Cost effectiveness
I. Drummond, M. F.
338.4'33621'0941   RA410.55.G7
ISBN 0-19-261398-7

Library of Congress Cataloging in Publication Data
(Revised for vol. 2)
Drummond, M. F.
Studies in economic appraisal in health care.
(Oxford medical publications)
Includes indexes.
1. Medical care—Cost effectiveness—Abstracts.
2. Medical economics—Abstracts.   I. Title.   II. Series.
RA410.5.D784   338.4'3621"0941   80-41312
ISBN 0-19-261398-7

Typeset by Latimer Trend and Company Ltd, Plymouth
Printed in Great Britain
at the University Printing House, Oxford
by David Stanford
Printer to the University

Preface to the second volume

Since the publication of the first volume of *Studies in economic appraisal in health care* there has been an increase in both the quantity and quality of published work in this field. Therefore there is a parallel increase in the need for users of the results of such appraisals to review them with a critical eye. We hope the new volume meets this need by reviewing a further 100 studies which, taken together, illustrate both the increased range of topics addressed and the methodological advances made.

Since, given the growth in the literature, it would be impossible for us to be comprehensive in our review, we have tried to draw out most of the general methodological points in the introduction to the volume (Chapter 2) and in the introductions to the six sections of the Summaries of Published Work. (In our view the introductions also have the effect of making the second volume of *Studies* a more readable and free-standing document.)

We have each taken responsibility for the production of a number of summaries, which have then been grouped together in the six sections. The sections have been edited by M.F.D. (sections 2, 3, 4, 5) and A. L. (sections 1, 6). The methodological introduction (Chapter 2) was written by M.F.D. and A.L. and the section introductions by M.F.D. (Sections 4 and 5), M.F.D. and K.L. (Section 2), A.L. (Sections 1 and 6) and A.L. and K.L. (Section 3). Within the constraints of time and logistics, we have also tried to offer constructive comments on each others' contributions. We hope that the result is a reasonably homogeneous and integrated volume.

As with the first volume of *Studies*, our purpose has not been to confront the authors of studies with their own shortcomings, nor to suggest that we could have done better. Rather it is to prevent others from accepting the methodology and results of published work without question and to stimulate them to seek improvements in the current state of the art.

Aberdeen
Birmingham
and
Manchester
November 1985

M.F.D.
A.L.
K.L.
A.S.

Preface to the first volume

The review and classification of studies contained in this volume is complementary to the methodological text contained in *Principles of economic appraisal in health care*. Where possible I have tried to indicate the places in the companion volume where more discussion can be obtained of the particular technical points arising from a review of published work in the field of economic appraisal in health care. Therefore I would expect that, apart from those very familiar with cost–benefit methodology, most readers will need to consult both volumes.

The review of the existing published work highlights the practical problems of undertaking economic appraisals in the field of health care. When considering my comments on existing studies, some of which are critical, it is essential to make allowances for the practical problems facing analysts. For example, from time to time I have criticized authors for omitting key categories of cost or benefit from their analysis. In most cases these omissions will be all too apparent to the individuals concerned, i.e. they would have included them if they could have calculated them! However, my purpose has not been to confront the authors concerned with their own shortcomings, nor to suggest that I could have done better. Rather it is to prevent others from accepting the methodology and results of published work without question and to stimulate them to seek improvements in the current state of the art.

Birmingham
February 1981   M.F.D.

Contents

Acknowledgements

A number of individuals have been particularly helpful in the production of this volume. Alan Williams and Ken Wright, the originators of the project that led to the production of the first volume of *Studies*, have continued to offer moral support. The main methodological contributions have come from Tony Culyer, Gavin Mooney, and Greg Stoddart, who have always been willing to offer constructive comments on drafts. A number of members of the Health Economists' Study Group, the Economic Advisers' Office of the DHSS, and colleagues (of M.F.D.) at McMaster University (Canada) greatly assisted us in locating the published studies. In addition, we have benefited throughout the project from excellent secretarial assistance at the Health Services Management Centre (Birmingham), the Health Economics Research Unit (Aberdeen), and the Department of Clinical Epidemiology and Biostatistics (McMaster).

Finally, we should like to thank our various 'significant others' who have given us unswerving support during the inordinate amount of time it has taken to bring this project to fruition.

Introduction

1.1. The scope and purpose of this volume

The first volume of *Studies in economic appraisal in health care* contained methodological critiques of 101 studies of the cost–benefit type published before 1980. It was intended to complement the methodological text *Principles in economic appraisal in health care* by highlighting the practical problems of undertaking economic appraisals in the health care field. The objective was not to confront study authors with their own shortcomings, but to prevent others from accepting the methodology and results of published work without question and to stimulate them to seek improvements in the current state of the art.

Since the publication of the first volume of studies there have been three changes. First, the quantity of published work has increased dramatically in more recent years. Secondly, as a consequence of the overall growth in the literature, there are now studies addressing a wider range of health care choices; hence the existing literature is now of interest to a wider range of health care professionals and other non-economists. Finally, there have been a number of methodological advances, partly as a consequence of economic analysts tackling a wider range of practical problems of resource allocation in the health care field.

Therefore, this second volume of *Studies in economic appraisal in health care* is intended to update the reader not only in terms of the range of applications of the cost–benefit approach, but also in terms of its methodological development. The basic method of reviewing and classifying studies remains the same as in the first volume (see below) and we expect therefore that the reader will consult *Principles in economic appraisal in health care* for exposition of the key methodological points. However, there are two innovations in this volume, which together we hope provide a response to the changes in the composition of the literature and also make this volume a more free-standing document.

First, the volume contains a methodological review of the studies as a group, entitled 'Current methodological issues in economic appraisal in health care'. This highlights the methodological developments exemplifed by the more recent studies and pinpoints recurring methodological

1

weaknesses. (In essence it constitutes a methodological apppendix to the companion volume *Principles in economic appraisal in health care*.) Secondly, the volume is sectioned in order to facilitate discussion, in a series of short review papers, of sub-groups of studies dealing with a common issue. The issues around which the studies are grouped are:

— burden of disease and alternatives in public policy
— alternatives in prevention
— alternatives in diagnosis
— alternatives in therapy
— alternative locations of care
— alternatives in health service organization

It is anticipated that such a grouping of the material will assist the reader wishing to compare and contrast a number of studies dealing with the same issue.

1.2. The classification scheme

For each study, comments are given under five main headings. These are: study design; assessment of costs and benefits; allowance for differential timing and uncertainty; results and conclusions; and general comments.

1.2.1. *Study design*

The comments under this heading relate to the points discussed in Chapter 3 of the companion volume. The question tackled by the study will be identified and placed into one of four broad categories.* These are:

(a) What is the cost of treatment?
(b) What is the benefit from treatment?
(c) What is the most efficient way to treat a given condition?
(d) Is the treatment worthwhile?

In addition, the alternatives appraised by the study will be stated. Finally, some general comments about the study design will be made. Were any relevant alternatives omitted? Did the study allude to the notion of the margin discussed in Section 3.2 of the companion volume? Was there anything in the design of the study that was likely to lead to confusion—such as the designation of averted costs as 'benefits'?

---

* In each study summary, the statement of the study question will be followed by the letters (a), (b), (c) or (d).

1.2.2. *Assessment of costs and benefits*

This element in the classification scheme relates to the material discussed in Chapter 4 of the companion volume. In the enumeration of the relevant costs and benefits, were any important items omitted? (The relevant costs and benefits are those relating to (*a*) the changes in resource use in the health sector and other sectors, (*b*) the changes in productive output, and (*c*) the changes in health state *per se*—see Fig. 4.2 of the companion volume, reproduced as Fig. 1.1.)

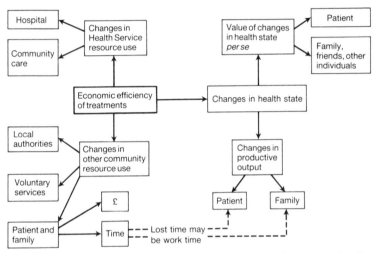

Fig. 1.1. The relevant changes in a comparison of the economic efficiency of treatments. *Source*: Drummond (1980).

In discussing the measurement of the costs and benefits employed in the study, particular attention will be paid to the problems of joint costs, and to the source of the underlying technical data. Was the economic appraisal based on proper medical, or other technical, evaluation? To what lengths did the authors of the study go in order to identify the precise changes (say) in resource utilization arising from the treatments being appraised? On this latter point it is often the case that insufficient details of the costing methods are given in the published study, so comments are likely to be limited. However, where the authors of a study clearly recognize the importance of identifying marginal (or incremental) costs, this fact will be noted.

With regard to the explicit valuation of costs and benefits, particular

3

note will be taken of the source of the values used. (See Section 4.3 of the companion volume.)

1.2.3. *Allowance for differential timing and uncertainty*

Adjustments for the differential timing of, or uncertainty in, costs and benefits will be noted. Were costs and benefits discounted; if so, at what discount rate? Was a sensitivity analysis performed? Were either of these two procedures overlooked in situations where they should have been employed? (See Chapter 5 of the companion volume.)

1.2.4. *Results and conclusions*

The results of the study will be recorded and, if the study reaches any firm conclusions, these will be noted. Study results are quoted mainly in order to give the reader a feel for the type of information typically produced by economic appraisals. Obviously, the applicability of the results and conclusions of a given study to one's own situation will depend upon local circumstances. Also, it should be remembered that the results are a product of the study methodology, and it would be advisable to consult the comments on this as well as the results themselves. If the study employs a decision rule this will be noted. There are a number of points to note in the use of decision rules and these are discussed in Chapter 6 of the companion volume. Occasionally results will be quoted directly from the study and will be placed within quotation marks.

1.2.5. *General comments*

This section in the classification will be used to accommodate points which do not fit easily into the other sections and to highlight important features of particular studies. For example, if a study is particularly strong in certain aspects of methodology, or breaks new methodological ground, this will be noted so that the reader is guided to the study as an example of how the methodology should be applied.

1.3. Criteria for inclusion in the volume

The criteria used were:

(i) The study should be concerned with the economic appraisal of alternative health care programmes.
(ii) The study should use real (as opposed to illustrative) data.
(iii) The study should be published (or easily obtainable) in the English language.

In practice these criteria proved workable, and have generated in excess of 100 studies. From these we have selected a group of studies which, taken together, illustrate the advances in this field since the publication of the first volume of *Studies in economic appraisal in health care* (Drummond 1981).

References

Drummond, M. F. (1980). *Principles of economic appraisal in health care.* Oxford University Press.

—— (1981) *Studies in economic appraisal in health care.* Oxford University Press.

Current methodological issues in economic appraisal in health care

The literature on economic appraisal in health care has grown rapidly over the last few years; not only has the quantity of published work increased (Warner and Luce 1982) but there have also been a number of methodological advances. The purpose of this chapter is to bring the reader up-to-date on current methodological issues in this field. A number of important methodological contributions will be highlighted, with examples, and recurring errors or deficiencies in approach pointed out.

At the same time, the comments here will provide guidance to the reader in assessing the quality of studies and indicate the elements which the individual study summaries attempt to highlight. This review is based primarily on studies summarized in this volume, most of which were published in the literature after 1978. Earlier studies, and the implicit state of the art at that time, were reviewed in the earlier volume of *Studies in economic appraisal in health care* (Drummond 1981a) and the associated methodological text, *Principles of economic appraisal in health care* (Drummond 1980).

The organization of the chapter follows the critical appraisal format used in Drummond (1981a) and in the summaries in this volume. That is, comments on study methodology will be grouped under the following main headings: study design, assessment of costs and benefits, allowance for differential timing and uncertainty (in costs and benefits) and presentation of results and conclusions.

2.1. Study design

2.1.1. *The viewpoint(s) for economic appraisal*

The viewpoint from which an economic appraisal is conducted is a key determinant of the range of costs and benefits considered. Economists argue for a societal viewpoint, which implies that the relevant range of costs and benefits is that set out in Fig. 1.1. There are, of course, other viewpoints, such as those of the government, providers, and patients. In the more recent literature it is much more common to find the viewpoint for analysis stated explicitly. Often this may not be the societal viewpoint; many studies investigate costs and benefits from the point of view of the particular Ministry of Health (see, for example, Hull *et al.* 1981).

Whilst the societal viewpoint is the most relevant, it is also of interest to investigate other viewpoints, because this:

—helps point out to any particular clients for the analysis (such as the government) how their own interests fit into the broader societal perspective;
—helps identify situations where individual viewpoints are at odds with the societal viewpoint. If, for example, providers were likely to be adversely affected by a new programme, this may lead to resistance to implementation of the socially preferred solution. Identification of such situations is the first step in developing a strategy for bringing about change.

2.1.2. *Use of decision trees or algorithms*

Many choices between alternative health treatments involve not just one decision, but a *sequence* of choices. For example, therapeutic decisions may depend on the prior decision of whether or not to perform a particular diagnostic test, and the results of that test. In an edited collection by Bunker *et al.* (1977), a number of applications of *decision analysis* in the surgical field were discussed. Since that time there have been applications in other fields of medicine and at least one general text (Weinstein and Fineberg 1980). The main contribution of decision analysis in the context of economic appraisal is that it provides a logical method of setting out a sequence of clinical decisions and the associated outcome probabilities. Obviously, this is an important prerequisite for assessing the costs and benefits of strategies. Apart from giving a framework for estimating costs and benefits, decision analysis can be useful in 'thinking through the problem' in a systematic manner. One powerful example of this is presented in the paper by Neuhauser (1978) on routine paediatric preoperative chest X-rays, where the author shows how decision analysis can be used to structure a clinical decision making problem. Some of the data used are merely illustrative and no attempt is made to justify them or to check their accuracy. However, the author points out that progress can still be made without precise estimation of all the parameters.

Studies in Economic Appraisal in Health Care

2.1.3. *Economic appraisal and controlled clinical trials*

It was argued in Drummond (1980) that economic appraisal should be based on reliable medical evidence, such as that produced by controlled clinical trials (preferably those involving random allocation of subjects to the study and control groups). The earlier critical review of the literature (Drummond 1981*a*) contained a number of examples of economic appraisals undertaken alongside clinical trials. (See, for example, Russell *et al.* 1977; Waller *et al.* 1978.) Issues surrounding the incorporation of economic analysis in clinical trials have recently been discussed by Drummond and Stoddart (1984). The main arguments for an economic component in clinical trials are that:

—some information collected during controlled trials is important for subsequent economic analysis. It would therefore be inconvenient or difficult to extract these data retrospectively. The types of data being referred to here include hospital length of stay, patients' costs and health status information. (See, for example, Culyer *et al.* 1983);

—for balanced decision making it is important to generate cost information at the same time as effectiveness information;

—inter-disciplinary work (between economists, clinicians, statisticians and epidemiologists) is beneficial to all parties. (Research into health status indices is probably the best testimony to this.)

The literature after 1978 contains a higher proportion of economic appraisals carried out in association with clinical trials. See, for example, Logan *et al.* (1981) and Weisbrod *et al.* (1980). In our view this trend is to be encouraged, although the following points should be noted. First, Evans and Robinson (1980) argue that experimental activity may induce changes in the care delivery system, thereby producing (economic) results atypical of a regularly operating system. For example, medical research protocols may require modifications in clinical procedures, or procedures may generally be carried out more meticulously during a period when research is being conducted. Therefore, it may not always be appropriate to undertake the economic study during the period of a randomized controlled trial. Certainly there is a need for more clinical trials of a 'management' or 'service' orientation, that try to assess therapies as delivered in an actual service setting. In addition, it is important that the economic analyst interprets the results obtained during an experimental period intelligently, especially when extrapolating to other settings. (This issue is further discussed in Section 2.4, below.)

Secondly, it could be argued that while it is beneficial for economic appraisal to be carried out in association with clinical trials where circumstances permit, there may be *additional* questions that can only be answered outside the context of clinical trials. Consider, for example, the economic assessment of a new medical technology such as cimetidine, an H_2 receptor antagonist which early clinical evaluation suggests may be effective in the treatment of peptic ulcer disease. One approach to economic evaluation of the technology would be closely integrated with clinical evaluation; the object would be to undertake economic analysis in association with controlled clinical trials of cimetidine *versus* alternatives (e.g. proximal vagotomy) for a range of clinical indications. The paper by Culyer and Maynard (1981) is an example of this approach, although it draws on medical evidence from a number of sources. However, it may also be considered relevant to estimate the *total economic impact* (in terms of costs and benefits) of the new technology. Such a study would need to consider not only the use of cimetidine in 'appropriate' clinical situations, where the patient has a confirmed ulcer and where clinical trials show that the new technology is effective. It would also need to consider the use of cimetidine in 'inappropriate' situations.

Finally, the *quality* of clinical trials in some of the recent studies still leaves room for improvement. The main problems which arise are faults in the design of the trial and insufficient numbers of patients to provide reliable results. For example, in their study of treatments for acute variceal bleeding, O'Donnell *et al.* (1980) studied only 32 patients although it was their intention to assess four treatment alternatives. The patients were not randomized between therapies nor were they matched in any way. (Detailed appraisal of the quality of medical evidence is outside the scope of this volume. For more discussion see Department of Clinical Epidemiology and Biostatistics 1981.)

2.2. Assessment of costs and benefits

2.2.1. *Health service resource use*

Many health treatments involve a period of hospitalization and it is the estimation of these costs that continue to present difficulties. Many studies still use *per diem* charges, or average hospital cost per day, as a basis for the calculation. This approach has a number of faults.

First, the average cost per day may not reflect the costs of the particular treatment or department that is the subject of interest. However, the estimation of the costs attributable to a particular category of patients is difficult since so many hospital resources are used jointly with other patients. Some progress is being made on this front. A minority of studies, such as that by Boyle *et al*. (1983), employ quite sophisticated methods of cost allocation to individual hospital departments. Others, such as that by Hull *et al*. (1982), separately identify those medical costs clearly attributable to the treatment of interest, allocate other 'hotel' costs on the basis of hospital average daily costs, and then test the sensitivity of the study result to the assumptions made (e.g. if the hospitalization costs were 50 per cent higher than assumed would this make any difference to the study conclusion?). The choice between these two approaches will depend on the importance of these costs in the study as a whole.

Secondly, the average daily hospital costs are likely to be a poor guide to the savings in resource use if hospital stays were shortened. This is because resource use is typically higher in the early part of a patient's hospital stay. Even in the literature since 1978, few studies address this *marginal costing* issue. Even fewer studies address the related issue, of how (or, indeed, *whether*) any freed resources are redeployed (Stoddart 1982). Evans and Robinson (1980), in an analysis of surgical day care, point out that the potential cost savings have not been achieved and go on to discuss the ways in which incentives could be changed (within the Canadian health care system) in order to realize such savings. It would be advantageous if the authors of economic appraisals took the discussion further, in the presentation of their results, and did not stop at the point of identifying *potential* benefits. However, authors should not be unduly criticized for stopping at the prior stage in the analysis since, after all, the methods of redeploying freed resources should be the subject of a separate, subsequent, managerial debate. Nevertheless, the authors of studies could assist this debate by emphasizing the fact that study results may depend crucially on key actors changing their habits or practice styles, and by highlighting the role that the clients of the results need to play in achieving the projected 'savings' or other benefits.

Finally, the authors of many studies are unquestioning in their use of actual hospital charges in valuing resource inputs or savings, even if the changes in resource use attributable to a given treatment have been correctly estimated. For example, in a study of burns treatment in specialist centres *versus* general hospitals, Linn *et al*. (1979) used hospital charges as the basis for the cost comparison. No account was taken of the fact that specialist centres may charge higher fees for reasons unrelated to the resources used by particular patient groups.

In countries with a fee-for-service system it is quite common for authors to take hospital or physician billings as estimates of the costs of (say) medical procedures or diagnostic tests, without any discussion of the basis for calculation of the billing figures. For example, do the billing figures for diagnostic tests cover both the professional fee and the costs of supplies and reagents? Are they also intended to cover the capital costs in purchase of equipment? Do the billing figures reflect potential economies of scale? Obviously the answers to these questions crucially affect *the other* costs to be included in the appraisal, alongside the billing figures.

A related point, which is raised here but is relevant to all types of cost and benefit, is the issue of price differences for the same resources in different locations. General price differences between locations (e.g. the USA and UK) may not be a problem, but differences in *relative* prices may lead to different outcomes from the analysis. This was the case in a study of the cost effectiveness of alternatives for the diagnosis of deep-vein thrombosis (Hull *et al.* 1981). The relatively high price charged for venography in the USA (compared to that in Canada) suggested a different diagnostic strategy for the disease in the two countries.

The results of studies which report resource changes only in money terms must, therefore, be interpreted carefully when being transposed to other settings. In fact, there is a strong case for presenting costs not only in their financial amounts, but also in terms of the physical amounts of each input used (e.g. nurse hours or nurse visits, number and category of each diagnostic test, number of hospital inpatient days, etc.). Presentation of the 'ingredients' of each treatment programme would greatly aid the interpretation of the results of a given study in a different setting.

Although an increasing number of authors give a reasonably full presentation of costing methods, the general picture is one of an imbalance between the amount of space devoted (in a published article) to the clinical and epidemiological aspects of a given study and that devoted to the estimation of costs. The lack of details given on costing methodology is a continuing frustration to those, like ourselves, wishing to assess the methodological quality of economic appraisals.

2.2.2. *Estimation of patient, family, and volunteer costs*

Most economic appraisals in health care continue to place more emphasis on health service costs and benefits than those falling on other parties. The reasons for this are twofold. First, the viewpoint of the Ministry of Health is often prominent; secondly, since prices are not readily observed for patient and volunteer time inputs, there are estimation difficulties.

Some of the studies reviewed in Drummond (1981a) imputed a value to patients' time (e.g. Schweitzer 1974). Very few studies attempted any detailed estimation of patient or family costs; the study by Waller *et al.* (1978) being an exception. The more recent literature offers few new

11

insights, although Evans and Robinson (1980) suggest that in some cases it would not be necessary to measure patients' costs. For example, if patients are offered a choice of therapy (say, day case surgery *versus* traditional inpatient methods), their revealed preference provides an answer to the question concerning their perceptions of the relative costs and benefits. If the patient's preference is for the procedure that is less costly to the rest of society, we need look no further.

In many cases the inclusion of patient and family costs would merely confirm the existing study result. However, the issue is considerably more complex in the case of (say) home care for the elderly or chronically sick. Here considerable burden may be placed on the family. On the other hand, caring for a sick relative can be very rewarding. The current literature contains very little empirical work in this area, although Fenton *et al.* (1982) test for statistically significant differences in categories of cost borne by families supporting relatives on a community-orientated mental illness programme *versus* a more traditional hospital based regimen. In our view the area of patient and family costs is one where both conceptual and empirical advances still remain to be made.

Finally, few studies in the recent literature consider the opportunity cost of volunteer time. Valuation of this poses the same kinds of conceptual and empirical challenges as for patient and family costs. For example, if a volunteer will only deliver meals-on-wheels, the cost to that programme is zero and the social opportunity cost may be negligible or zero. If the volunteer will work on a number of tasks, although the cost to a specific programme is zero, the social opportunity cost depends on the value of the volunteer in those alternative uses. If one is working on a specific appraisal it will be exceedingly difficult to work out all the social opportunity costs for all volunteers in a particular locality. However, it is surprising that evaluations of new programmes involving substantial amounts of volunteer time do not investigate whether volunteers have been diverted from other programmes, or whether the total pool of volunteer time has been increased (either by more volunteers coming forward or by existing volunteers giving up even more of their leisure time). Nevertheless, one should not understate the estimation problems.

Although evaluation from the societal perspective would consider volunteer costs irrespective of the source of the volunteers, policy makers might be particularly concerned with the impact on the availability of volunteers for other public programmes.

2.2.3. *Consideration of gains/losses in productive output*

In the literature published before 1978, it was common for analysts to argue that the potential to avert production losses provides a major justification for investment in health services. (See, for example, Longmore and Rehann 1977.) This is no longer the case; the main debate centres on *whether they should be included at all*, as *part* of the benefit of health service investments (or conversely, as *part* of the cost of giving treatment) (Drummond 1981*b*).

The main argument *for* their inclusion is that production gains or losses are important to the community; other things being equal we would prefer a health treatment that removed a patient from the workforce for a shorter time. Also, in the long run production losses affect the community's ability to maintain its level of provision of goods and services. On the other hand, the critics of their inclusion point out that:

—earnings losses (the valuation method used) are a poor measure of production losses;

—it is not clear that production *is* actually lost, especially in times of unemployment;

—society's willingness-to-pay for averting lost production may have been already incorporated in other estimates of health benefits (e.g. patients' utility or willingness-to-pay estimates);

—incorporation of production losses into cost and benefit estimates leads to priority setting (between treatments and conditions) based on the current job productivity of individuals suffering from the diseases in question. Health policy makers may not want to set priorities in this way.

On the issue of whether or not production is lost, two cases need to be distinguished; individual short-term illness and the level of ill-health in the community at large. In the case of short-term illness affecting one individual, there may be no noticeable change in a firm's production. Staffing levels within large organizations are often based on the assumption of a level of sickness absence across the total pool of employees. However, a reduction in the level of sickness absence would release workers for other activities, but this would not necessarily result in a proportionate change in the firm's production.

This raises the second aspect of this debate; whether or not additions to the total labour force will be productively employed. Holtermann and Burchell (1981) argue that it is sufficient to adjust for the prevailing rate of unemployment. This view is based on the assumption that an objective of macro-economic policy is to maintain a given rate of unemployment. Demand management will ensure that additions to the workforce are absorbed, up to this rate of unemployment. Although this may be an objective of macro-economic policy, it may not be achieved, however. Therefore, the prevailing macro-economic policy in a given country should certainly be taken into account when undertaking one's study and assumptions about this policy subjected to sensitivity analysis.

The conflict concerning the influence of job productivity on priority setting arises from the attempts to integrate, in health care evaluations, two valuation systems. This problem has been described by Williams (1981) as 'mixing sugar and sand'. He points out that in the non-health sector we have valuations 'contaminated' by unequal ability to pay, whereas in the health sector we want a set of 'decontaminated' values. But which of these should we use to measure the opportunity cost of the resources used in the health care sector? Williams argues that as long as society thinks it right that non-health benefits should be valued, relative to one another, by the 'contaminated' set of values, then there are few strong arguments for excluding the production benefits, once benefits in the form of savings of health service costs are included in the analysis.

There is no easy way out of this dilemma, except to be clear in one's analysis what is being included and why, and to point out to decision makers the implications of adopting the study results.

2.2.4. *Valuation of health states*

At the time of the earlier critical review (Drummond 1980, 1981*a*) it was argued that the relative valuation of health states was 'still largely experimental in its nature'. Since that time there has been a growth in the number of studies expressing benefits in 'quality adjusted life years', and accompanying advances in the measurement techniques. It has become customary to refer to these studies as *cost–utility analyses*, reflecting the fact that patients' utilities or valuations of health states are explored.

The growth in this literature arises partly from dissatisfactions, expressed by a number of authors, with cost–benefit analysis *as applied in practice*. (See, for example, Drummond 1981*b*; Klarman 1982.) The problem is that most cost–benefit analyses concentrate on easily measurable aspects of health benefits, such as averted production losses. Therefore, it might be preferable to value health states *relative to one another*, than to attempt money valuations which vary widely depending on the method used. In theory, if health status indices could be developed to allow explicit inter-programme evaluation, then the only additional question answered by valuing health outputs in money terms would be that of how large in total the health care sector budget should be.

Culyer (1978) argued that 'at the present state of development of the art, it does not seem realistic to require that these different measures should be made comparable with one another for the purposes of explicit inter-programme evaluation' and that their use is seen as being in evaluating alternative ways of providing a specific service. However, in a recent paper Kaplan and Bush (1982) suggest that a 'cost–utility ratio can be used in a general health policy model to compare the efficiency of different programmes or to assess the relative contribution of different programmes and providers in the health care system'. In the same paper they go on to present ratios (in cost per well year) for a diverse set of programmes, as calculated in earlier studies. They propose (as a basis for debate) that programmes with ratios of less than US$20 000 per well year would be cost-effective by current standards, whereas those with ratios greater than US$100 000 per well year may be questionable in comparison with other health care expenditures. (Programmes with ratios between US$20 000 and US$100 000 per well year were thought to be controversial, but probably justifiable by many current examples.)

Whilst many might consider the approach of Kaplan and Bush (1982) to be at best heroic, it is clear that research into the reliability, comparability and policy relevance of health status indices should be carried out, in order that the returns from the investment in the development of these measures can be harvested.

Another feature of the health status valuation literature is the growth of two distinct approaches. The first approach, outlined by Torrance (1982), places emphasis on the development of measurement approaches and empirical testing on different populations. The other approach, outlined by Weinstein (1981), places emphasis on the estimation of utility values by a quick (and inexpensive) consensus-forming exercise and then the undertaking of extensive sensitivity analysis on the chosen values to see whether study results change if the chosen utility estimates are varied. We see a role for both approaches. The latter approach is useful in getting decision makers to think about resource allocation problems and is relatively quick and inexpensive. The measurement approach is useful in highlighting the fact that different actors (doctors, policy makers, patients and the general public) may have different values and is clearly crucial in situations where the study result *is* sensitive to the utility values assigned. (An example of such a case arose in the study by Stason and Weinstein 1977, on the economics of hypertension therapy. The study result was sensitive to whether it was assumed that the side-effects of anti-hypertensive drugs constituted a 1 or 2 per cent reduction in health status.)

15

2.3. Allowance for differential timing and uncertainty in costs and benefits

2.3.1. *Allowance for differential timing*

The general justification for this was outlined in Drummond (1980), together with examples of the discounting calculation. The principle of discounting (to present values) is now widely observed in the health care evaluation literature but the following points, based on consideration of the more recent studies, are worth noting.

Despite disagreement amongst economists concerning the choice of discount rate, evaluation *practice* seems to be coverging in that typically rates between 2 and 10 per cent (in real terms) are chosen and a sensitivity analysis performed. The UK government currently advises a real rate of 5 per cent for health service projects (H.M. Treasury 1982). Often it can be shown that the choice of rate within the range of 2–10 per cent does not affect the study result (Lowson *et al.* 1981).

Secondly, while almost everyone accepts that costs should be discounted, there is still disquiet amongst non-economists over the discounting of effects such as improved health status. The arguments typically put forward are that health benefits (e.g. years of life) are not reinvestable in the way that cash flows are and that lives saved at different points in time not the same persons' lives; it is therefore sometimes argued that it is inequitable to weight these differently by applying the discounting procedure.

However, there are a number of strong counter-arguments and these include the following:

— the discussion can be recouched in terms of risk reduction rather than life years (Mooney 1977). Not only does this make the argument less emotive, but also highlights the fact that, for the individual, the benefit of reducing risk of death in the future is dependent upon the probability of being alive in the future. Therefore, a reduction in risk of death *now* will have a higher value than some future risk reduction yielding the same increase in life years;
— pure time preference (on the part of individuals or the community) is not necessarily to do with the scope for reinvestment. In any case it is possible to think of individuals trading the *quality* of life through time; that is, making sacrifices now in return for healthy time later;
— the 'inter-generational equity' issue can be addressed independently of discounting;
— it is asymmetrical to discount costs but not effects. If project A beginning in year one has the same costs and benefits as project B

beginning in year 10, project B would always be preferred to project A (i.e. the investment postponed) if costs were discounted and benefits not;

—lack of discounting can give rise to quite impossible results, e.g. a health programme which saves one life each and every year (forever) would be worthwhile whatever the size of the initial investment.

Finally, it is interesting to note that while the earlier economic appraisal literature in health care included many examples of analysts discounting costs to present values, there are more studies in the current literature which employ the same principles in reverse; namely the conversion of capital outlays to *equivalent annual costs*. This approach has a number of attractions. First, it may be more familiar to most people, being the principle upon which house mortgage repayments are based. Secondly, the process of annuitization is equivalent to imputing an annual rental value for a piece of equipment or a building and this provides a more intuitive explanation for non-economists. Finally, it is sometimes a more convenient approach, since most health treatment expenditures are on an annual, recurring basis. See Lowson *et al.* (1981) for a recent application.

2.3.2. *Allowance for uncertainty*

It was pointed out in Drummond (1980) that there are a number of sources of uncertainty in cost and benefit estimates. Two remedies were suggested: (*i*) to obtain better estimates of the magnitudes or values of the items (by basing economic appraisals more frequently on controlled clinical trials, for example), and (*ii*) to undertake a sensitivity analysis of study results to variations in the factors which remain in doubt.

The improvement in the literature on the first count has already been mentioned. There is improvement on the second count too, in that it is now most unusual to find studies without some degree of sensitivity analysis. However, a number of studies could still be considered to be methodologically unsound, in that not enough attention is paid to the shape of the probability distribution of the variables in question, or that not enough justification is given for the choice of variables incorporated in the sensitivity analysis and the range of values considered.

Finally, it is worth highlighting another practice which has gained more prominence since 1978. It is now fairly common to see tests of statistical significance performed on observed cost differences between treatment alternatives. (See, for example, Logan *et al.* 1981, and Fenton *et al.* 1982.) This practice is, of course, primarily a result of infiltration into economic appraisal by medical statisticians, as part of multidiscip-

17

linary activity in this area. It is to be welcomed; statistical tests are particularly useful and valid where the cost estimates obtained are the result of observation of individual patients receiving the alternative procedures. For example, lengths of stay of patients undergoing 'short-stay' and 'long-stay' surgery, or consumption of nursing and other services by patients undergoing 'institutional' and 'community-orientated' treatments for mental illness. However, many of the costs presented in economic studies are not estimated on the basis of an observed distribution, but reflect a protocol for a new procedure (which has not yet been used), or a consideration of the options facing the decision maker (e.g. can a hospital ward be closed, thereby freeing resources for other uses?). In such cases it would usually be acceptable to undertake a sensitivity analysis around a single figure, or to undertake an expected value calculation.

2.4 Presentation of results and conclusions

2.4.1. *Use of decision indices*

A number of common decision indices were outlined in Drummond (1980), where it was pointed out that these should be seen merely as a way of summarizing the results of an economic appraisal and that there are dangers in calculating them when they do not encompass all the relevant information.

In the recent literature it is much more common to observe the use of *incremental* cost-effectiveness ratios. This is important as, when evaluating a given health care programme or treatment, a comparison must always be made with a relevant alternative. If an analyst reports a cost effectiveness ratio for a single programme this is likely to reflect a poor understanding of appraisal methods unless it is the case both that the relevant alternative is 'to do nothing' *and* under the 'do nothing' alternative there will be no costs and no effects. These conditions rarely hold so the relevant ratio to report is normally the incremental cost-effectiveness ratio (of a given programme over and above the most relevant alternative). A number of studies in the more recent literature use this approach.

A comparison of diagnostic strategies for deep-vein thrombosis (Hull *et al.* 1981) is set out in Table 2.1. The cost effectiveness ratios for the three strategies are quite similar: Canadian $2784, $2989 and $3908 (per case detected) respectively. However, as the authors point out, the correct way to interpret these data would be to consider incremental cost-effectiveness. In this case the incremental cost-effectiveness ratio of adding leg scanning to impedance plethysmography (IPG) is $3683, whereas a change to venography would have an incremental cost-effectiveness ratio of $13 852, compared to IPG plus leg scanning. Of course, there are many pairways comparisons to be made and in this particular case the authors considered a number of these.

Table 2.1 Cost-effectiveness of diagnostic strategies for deep-vein thrombosis (after Hull *et al.* 1981)

Diagnostic procedure	No. of patients diagnsed (out of 201)	Costs ($)	Cost per patient diagnosed ($)	Incremental cost per additional patient diagnosed	
				over IPG alone	over IPG + leg scanning
Impedance plethysmography (IPG) alone	142	395 359	2784	–	–
IPG plus leg scanning	184	550 046	2989	3683	–
Venography	201	785 538	3908	6613	13 852

2.4.2. *Indicating the relevance of study results to clinical or planning decisions*

In the earlier literature reviewed in Drummond (1981*a*), few studies devoted much space to this topic. It has two dimensions, namely (*i*) selection of alternatives that are clinically or managerially relevant, and (*ii*) discussion of how study results should be interpreted in the light of different clinical or planning circumstances.

With regard to the selection of relevant alternatives, recently there have been many more studies exploring the economics of basing clinical decisions on the presence or absence of particular clinical indications. For example, Holmin *et al.* (1980) investigated the cost-effectiveness of *routine* intra-operative cholangiography *versus selective* use of the procedure, on patients who fulfilled at least one of four common clinical criteria (colic attacks, cholecystitis, pancreatitis, and jaundice).

With regard to the interpretation of study results in the light of one's own circumstances, the study by Lowson *et al.* (1981) on alternative modes of delivering long-term oxygen therapy to chronic bronchitics explores the relationship of costs to size of patient population served (in a given locality) and the availability of existing workshop facilities for the maintenance of oxygen concentrators.

2.4.3. *Consideration of factors other than economic efficiency*

The assessment of the economic efficiency of health care alternatives is the main concern of economic appraisal. However, it is often useful, in the discussion of results, to allude to other factors relevant in making decisions about the allocation of health care resources.

For example, the distributive aspects of options (i.e. equity) are important as well as their efficiency. Glass (1979) gives an excellent description of the choice (in screening for asymptomatic bacteriuria in schoolchildren) between a more cost-effective unsupervised test and a supervised test which results in a higher yield among children from lower-income families. The more recent batch of literature contains very few studies where the trade-offs between equity and efficiency are explored to any large extent. In fact there are one or two studies where the discussion is severely lacking in this respect. For example, in an analysis of the economics of measles immunization in Zambia, Ponnig-haus (1980) argues that the pay-off (in terms of health service resource savings) would be greater in urban areas because the utilization of services would fall. Of course, the rural area programmes have no chance of showing this category of benefit, as the same infrastructure of health services does not exist. Although other parts of the discussion in Ponnighaus' paper allude to this fact, simple reading of the study may lead the decision maker to think that it provides evidence for more investment in urban areas.

The other major factor in decision making, often overlooked by economic analysis, is the cost of implementing the preferred course of action. As was pointed out earlier, economic appraisal can identify the *potential* for efficiency, but some managerial or organizational changes may be so costly that any efficiency gains are wiped out. The more recent literature does not address this point very thoroughly, although it is encouraging to see a number of studies addressing managerial (as opposed to clinical) alternatives. (See, for example, the study by Ruchlin *et al.* 1982, on second opinion consultation programmes in surgery.)

Of course, a linked issue is that of the incentives, to the various key actors, required to bring about a change to a more efficient procedure. This brings the debate back to our starting point, that of viewpoints for economic appraisal.

20

2.5. Concluding remarks

In this chapter we have discussed the current state of the art in economic appraisal in health care, based on the studies summarized in this volume. Our review shows that progress has been made in a number of areas since the publication of the earlier volume of *Studies in economic appraisal in health care*. For more discussion of the methodological features of the individual studies, the reader should now turn to the summaries of published work contained in the next section.

References

Boyle, M. H., Torrance, G. W., Sinclair, J. C. and Horwood, S. P. (1983). Economic evaluation of neonatal intensive care of very-low-birth-weight infants. *New England Journal of Medicine* **308**, 1330–7.

Bunker, J. P., Barnes, B. A. and Mosteller, F. (1977). *Costs, risks and benefits of surgery*. Oxford University Press, New York.

Culyer, A. J. (1978). *Measuring health: lessons for Ontario*. Toronto University Press.

—— and Maynard, A. K. (1981). Cost-effectiveness of duodenal ulcer treatment. *Social Science and Medicine* **15c**, 3–11.

——, Macfie, J. and Wagstaffe, A. (1983). Cost-effectiveness of foam elastomer and gauze dressings in the management of open perineal wounds. *Social Science and Medicine* **17**, 1047–53.

Department of Clinical Epidemiology and Biostatistics (1981). How to read clinical journals. V. To distinguish useful from useless or even harmful therapy. *Canadian Medical Association Journal* **124**, 1256–62.

Drummond, M. F. (1980). *Principles of economic appraisal in health care*. Oxford University Press.

—— (1981a). *Studies in economic appraisal in health care*. Oxford University Press.

—— (1981b). Welfare economics and cost benefit analysis in health care. *Scottish Journal of Political Economy* **28**, 125–45.

—— and Stoddart, G. L. (1984). Economic analysis and clinical trials. *Controlled Clinical Trials* **5**, 115–28.

Evans, R. G. and Robinson, G. C. (1980). Surgical day care: measurements of the economic payoff. *Canadian Medical Association Journal* **123**, 873–80.

Fenton, F. R., Tessier, L., Contandriopoulos, A.–P., Nguyer, H. and Struening, E. L. (1982). A comparative trial of home and hospital psychiatric treatment: financial costs. *Canadian Journal of Psychiatry* **27**(3), 177–87.

Studies in Economic Appraisal in Health Care

Glass, N. (1979). Evaluation of health services developments. In *Economics and health planning* (K. Lee, ed.). Croom Helm, London.

H.M. Treasury (1982). *Investment appraisal in the public sector.* HMSO, London.

Holmin, J., Jönsson, B., Lindgren, B., Olsson, S.-A., Peterson, B. G., Sörbris, R., and Bengmark, S. (1980). Selective or routine intraoperative cholangiography: a cost-effectiveness analysis. *World Journal of Surgery* **4**, 315–22.

Holtermann, S. and Burchell, A. (1981). The costs of alcohol misuse. *Government Economic Service Working Paper No. 37.* HMSO, London.

Hull, R., Hirsh, J., Sackett, D. L. and Stoddart, G. L. (1981). Cost effectiveness of clinical diagnosis, venography and non-invasive testing in patients with symptomatic deep-vein thrombosis. *New England Journal of Medicine* **304**, 1561–7.

——, ——, Sackett, D. L. and Stoddart, G. L. (1982). Cost effectiveness of primary and secondary prevention of fatal pulmonary embolism in high-risk surgical patients. *Canadian Medical Association Journal* **127**, 990–5.

Kaplan, R. M. and Bush, J. W. (1982). Health-related quality of life measurement for evaluation research and policy analysis. *Health Psychology* **1**(1), 61–80.

Klarman, H. E. (1982). The road to cost-effectiveness analysis. *Milbank Memorial Fund Quarterly* **60**(4), 585–603.

Linn, B. S., Stephenson, S. E., Bergstresser, P. and Smith, J. (1979). Do dollars spent relate to outcomes in burn care? *Medical Care* **17**, 835–43.

Logan, A. G., Milne, B. J., Achber, C., Campbell, W. P. and Haynes, R. B. (1981). Cost-effectiveness of a worksite hypertension treatment program. *Hypertension* **3**(2), 211–18.

Longmore, D. B. and Rehann, M. (1975). The cumulative cost of death. *The Lancet* **i**, 1023–5.

Lowson, K. V., Drummond, M. F. and Bishop, J. M. (1981). Costing new services: long term domiciliary oxygen therapy. *Lancet* **i**, 1146–9.

Mooney, G. (1977). *The valuation of human life.* Macmillan, London.

Neuhauser, D. (1978). Cost effective clinical decision making: are routine paediatric preoperative chest x-rays worth it? *Annales de Radiologie* **21**, 80–3.

O'Donnell, T. F., Gembarowicz, R. M., Callow, A. D., Panker, S. C., Kelly, J. J. and Deterling, R. A. (1980). The economic impact of acute variceal bleeding: cost-effectiveness implications for medical and surgical therapy. *Surgery* **88**, 693–701.

Ponnighaus, J. M. (1980). The cost/benefit of measles immunization: a study from Southern Zambia. *Journal of Tropical Medicine and Hygiene* **83**, 141–9.

Ruchlin, H. S., Finkel, M. L. and McCarthy E. G. (1982). The efficacy of second-opinion consultation programs: a cost–benefit perspective. *Medical Care* **20(1)**, 3–20.

Russell, I. T., Devlin, H. B., Fell, M., Glass, N. J. and Newell, D. J. (1977). Day-case surgery for hernias and haemorrhoids: a clinical, social and economic evaluation. *Lancet* **i**, 844–7.

Schweitzer, S. O. (1974). Cost-effectiveness of early detection of disease. *Health Services Research* **9**, 22–32.

Stason, W. B. and Weinstein, M. C. (1977). Allocation of resources to manage hypertension. *New England Journal of Medicine* **296**, 732–7.

Stoddart, G. L. (1982). Economic evaluation methods and health policy. *Evaluation and the Health Professions* **5**, 393–414.

Torrance, G. W. (1982). Preferences for health states: a review of measurement methods. In *Clinical and Economic Evaluation of Perinatal Programs* (J. C. Sinclair, ed.). Proceedings of Mead Johnson Symposium on Perinatal and Developmental Medicine, Number 20. Vail, Colorado, June 6–10, 1982, pp. 37–45.

Waller, J., Adler, M., Creese, A. and Thorne, S. (1978). *Early discharge from hospital for patients with hernia or varicose veins*. Department of Health and Social Security, HMSO, London.

Warner, K. E. and Luce, B. R. (1982). *Cost-benefit and cost-effectiveness analysis in health care*. Health Administration Press, Ann Arbor.

Weinstein, M. C. (1981). Economic assessments of medical practices and technologies. *Medical Decision Making* **1(4)**, 309–30.

—— and Fineberg, H. V. (1980). *Clinical decision analysis*. W. B. Saunders, Philadelphia.

Weisbrod, B. A., Test, M. A. and Stein, L. I. (1980). Alternatives to mental hospital treatment: economic cost-benefit analysis. *Archives of General Psychiatry* **37**, 400–5.

Williams, A. H. (1981). Welfare economics and health status measurement. In *Health, economics and health economics* (J van der Gaag and M. Perlman eds.). North Holland, Amsterdam.

Summaries of published work

SECTION 1: Burden of disease and alternatives in public policy

Introduction

The 12 studies in this section are all concerned with some aspect of formation of public policy or priorities in health care. Most of the studies are primarily disease costing studies, that is they are concerned with quantifying the *potential* benefit from preventing all cases of a particular disease or health problem. (Bodkin *et al.* 1982; Harturian *et al.* 1980; Holtermann and Burchell 1981; Johnson and Heler 1983; Mills and Thompson 1978; Muller and Caton 1983; Yelin *et al.* 1979.) Lindgren (1981) has a wider focus; he estimates the cost associated with all disease and premature death in Sweden. Such studies can provide a starting point for the formulation of priorities between different health problems and some of the studies draw explicit comparisons between the burden of disease for different conditions (Harturian *et al.* 1980; Holtermann and Burchell 1981).

The four other studies in this section deal with different aspects of public policy decisions that fall outside the disease costing framework. The study by Muller (1980) is an example of the application of cost–benefit analysis to legislative intervention on health grounds (motorcycle helmet laws). The paper on heart transplants (Haberman 1980) is included here, rather than in Section 4, because it raises the wider issues of value of life. Also, in the UK context, the funding of this programme is very much a public policy issue. Chapalain (1978) assesses the priorities *within* a programme aimed at reducing perinatal mortality and morbidity. Cavenaugh (1981) is concerned with costing the effects of laboratory errors as an input to the formulation of policy to improve laboratory performance.

Particular methodological problems in this area

The two main issues arising from studies in this section relate to the valuation of lost production and the use of discounting. The loss of productive output is an important element in most burden of disease or

27

public policy studies and therefore it is essential that it is dealt with properly. However, this is frequently not the case. Quite apart from the general criticism of wages as a proxy for production losses, discussed in Chapter 2, a number of other problems arise. First and foremost, production losses should be proxied by *gross* employment costs, i.e. the total cost of the employee to the firm, since it is this amount that economic theory states should represent the value of the worker's output. Several studies refer to the use of wage rates or earnings without making it clear whether other costs, such as employer payroll taxes, are included. Holtermann and Burchell (1981) are an exception, in that they clearly state that gross employment costs have been used in their study.

This problem is compounded if employee taxes are netted out of earnings (Haberman 1980). This only indicates the financial loss to the employee but Haberman takes this to represent the contribution to the national economy. A further error in this particular study is the double counting of production loss and state benefits (transfer payments). Again, many of the studies in this section are not specific about whether or not transfer payments are included.

On the question of discounting, several studies avoid the difficulties by limiting their consideration of costs and benefits to a single year and this issue is taken up below. In other cases, it is not clear whether long-term consequences are included or if they have been properly discounted (Cavenaugh 1981; Chapalain 1978). The real problems, however, arise from the failure to distinguish between *discounting* costs or benefits and *deflating* prices. That is, there is confusion between the principle of reducing costs and benefits to their present value and the adjustment of different price levels to a common base. This confusion is evident in the study by Johnson and Heler (1983) where both discounting and deflation are carried out and referred to simply as 'discounting', whereas Haberman (1980) fails to discount at all because of the mistaken belief that this is required only to take account of price inflation. Another long-term consequence which most studies ignore is that preventing mortality from one disease will mean an increase in deaths from other disease, although occurring later. Death can only be delayed.

Current state of the art

One of the first things to notice about the studies in this section is that although reference may be made to economic costs, even in the study title, the range of costs actually considered is often quite limited. Many of the studies concentrate on easily quantified, financial costs or benefits. Whilst this approach is justified if, for example, a limited estimate of the

28

benefits is shown to exceed all the costs (Muller 1980), few of the studies address this issue.

Given the problems involved, it is understandable that not all studies will break new ground in measuring and valuing the intangible aspects of morbidity and mortality. However, almost all the studies reviewed here dismiss the problem early on and fail even to exploit work that has been carried out by others, for example, in valuing life years gained. Apart from the study by Chapalain (1978), none of the studies use any form of health status index. Even this example is open to criticism as it gives equal weight to morbidity and mortality in assessing perinatal outcomes.

The failure to measure or value health status *per se* leaves a very large gap, particularly in the burden of disease literature, and it is disappointing that studies in this area appear to have turned away from the problems posed by this issue. As a result, the priorities which may emerge from such studies will be distorted by the narrow definition of 'economic' costs and benefits adopted. This will, in its turn, exacerbate the tendency to criticize economic appraisal (rather than its improper use) for being biased, such as in favour of productive groups against the retired. Priorities between, for example, cancer and cardiovascular disease may also be distorted if the unpleasantness of the process of dying from the former is not taken into account. At a time when the more 'clinical' studies of choices in prevention, diagnosis or therapy have been developing the use of 'quality adjustment' to standardize the life years gained, the authors of burden of disease studies have not attempted similar work to improve comparability of their results with those of others.

Another important feature of the studies reviewed here is the extent to which available data are exploited rather than special data collected and 'expert judgement' used in place of direct observation. Whilst this can be a sensible strategy to adopt, particularly for studies mainly concerned with developing or demonstrating an approach to problems (Cavenaugh 1981), two points should be noted. First, there is an even greater need to test the sensitivity of the results against the data but in many instances this has not been done. Secondly, there is little evidence to show that exercises have been repeated with better data once the approach has been demonstrated to be successful or if the data available are shown to be inadequate.

Finally, a general problem with the 'cost of illness' studies is that the *objective* of the study is often unclear. As Holtermann and Burchell (1981) point out, 'a measure of total resource cost is of somewhat limited usefulness ... it would be better to study specific measures. ... and to compare the marginal savings produced with the extra cost of the measure themselves.' However, few of the studies reviewed here go further than measuring the burden of disease, presumably with the intention of influencing priorities. This is a somewhat circular argument since the resources already devoted to a particular disease are included in the cost of illness. Diseases which already have a lot of resources devoted to them may thus appear to be high priority cases for more resources.

Cost of illness studies can be used in a number of ways. Lindgren (1981) discusses some examples and the different methods of calculating the burden of disease that are appropriate. If burden of disease studies are not going to analyse proposals for research or prevention, they should at least present their information in the most useful way possible. In general, this will include distinguishing between direct costs, indirect costs, and intangible costs.

Contribution to decision making

Studies in this section certainly ought to be making a contribution to decision making as this is virtually their *raison de'être*. The studies fall into two categories in this context; those which look at the consequences of public decisions, such as the repeal of the compulsory wearing of motorcycle helmets in the United States (Muller 1980); and those which provide information to shape priorities within public policy, as do most of the burden of disease studies. However, it seems that such studies have had little impact, certainly in the UK on general policy formation, although studies on specific issues may have had a little more success.

The UK health departments have issued various documents and statements on priorities but there is little evidence in these of any influence from the approach of economic appraisal let alone the results of particular studies. However, the results of the study of priorities for reducing perinatal morbidity and mortality were implemented in France (Chapalain 1978) and the paper reports on the follow-up. This isolated success is probably due to the fact that the study also considered specific options for prevention, which most of the studies do not. The estimates of the costs of alcohol misuse (Holtermann and Burchell 1981) may have helped to shape policy in this area. It should be noted that in both of these examples, the authors worked within the respective health departments and, therefore, it is to be expected that they were working on problems already attracting the interest of policy makers. This may lend credence to the view that studies of burden of disease and policy priorities can be useful in decision making, provided that they tackle problems that policy makers are aware of, but that they are having less effect on what gets on to the policy agenda.

1 Bodkin, C. M., Pigott, J. J., and Mann, J. R. (1982). Financial burden of childhood cancer. *British Medical Journal* **284**, 1542–4.

Study design

1.1. *Study question*

What are the financial costs incurred by the families of children diagnosed as having cancer?(b)

1.2. *Alternatives appraised*

Existing situation *versus* hypothetical eradication.

1.3. *Comments*

This study was concerned specifically with estimating the financial burden of childhood cancer incurred during the first in-patient week of treatment and during a subsequent week of out-patient treatment, rather than during the complete episode of illness.

2. Assessment of costs and benefits

2.1. *Enumeration*

The financial costs considered comprised reductions in parental income and additional expenses incurred (travel, additional food, heating, clothing and presents) as a result of the child's illness.

2.2 *Measurement*

The families of 73 of the 98 newly diagnosed cases of childhood cancer at the regional oncology department in Birmingham were invited to participate in the study. The invitation to participate was issued on a non-selective basis.

Family information and details of costs incurred as a result of the child's cancer were collected by means of a questionnaire administered by a social worker. This information was collected during the first week following diagnosis and during the patient's treatment as an out-patient. In addition, a sample of 10 funerals was costed.

2.3. *Explicit valuation*

Presumably market prices were used, although details were not given.

3. Allowance for differential timing and uncertainty

Discounting was not relevant to the range of costs considered. In most cases, only the average results were reported and little information was given on the distribution of costs.

4. Results and conclusions

Of the 73 families invited to participate in the study, 59 were able to give the information required for estimation of family costs during inpatient treatment. Seven families could only provide partial responses and seven more declined to participate. With the exception of the lack of representation of social class I families, the study sample was representative of the childhood cancer population and families were of similar socio-economic status to the general population.

During the first in-patient week of treatment, the sum of income lost plus additional expenditure exceeded 50 per cent of total income in over 45 per cent of families; the range of loss of income experienced was £27.50 (the average for social class IV) to £6.38 (the average for social class V), and of increased expenditure £10.21 (the average for social class IV) to £19.77 (the average for social class II).

Twenty-two families provided information relating to reduced income and additional expenditure incurred during a week in which their child was receiving out-patient treatment. During this time, loss of income plus additional expenditure amounted to more than 20 per cent of income in over half the families. The greatest expenditure was incurred by families living furthest away from the oncology centre.

The average cost of the 10 funerals considered was £246 (range £87–£434).

5. General comments

This study investigated categories of cost which are often omitted from disease costing studies due to the difficulty of data collection. As the results indicated, costs falling on families may constitute a significant financial burden. In addition to the financial costs resulting from the child's cancer, parents may suffer considerable stress leading to the need for medical/psychiatric care, but such considerations were outside the scope of the study.

It is not possible from this study to estimate the financial burden to the family of the entire cancer episode, as no tests were made of the representativeness of the costs measured.

2 Cavenaugh, E. L. (1981). A method for determining costs associated with laboratory error. *American Journal of Public Health* **71**, 831–4.

1. Study design

1.1. *Study question*

What are the costs associated with laboratory error?(a)

1.2. *Alternatives appraised*

Laboratory error *versus* no laboratory error for three tests; prothrombin time, blood urea nitrogen and cholesterol.

1.3. *Comments*

The study develops a methodology for costing laboratory error using three tests as examples. The question is posed partly as a step towards considering what it is worthwhile spending to improve the quality of laboratory performance.

2. Assessment of costs and benefits

2.1 *Enumeration*

The author set out to consider costs that were easily quantifiable in dollar terms. It was not entirely clear from the paper which costs were finally included. Medical care costs were covered, but for other costs discussed (e.g. loss of livelihood or premature death) no details of any costing procedure were given. Laboratory errors were taken to be 'any laboratory test result which is different enough from true value to cause a change in the medical management of the patient'.

2.2. *Measurement*

The possible consequences of laboratory errors for the three tests were elicited from a panel of 20 physicians. These consequences ranged from the necessity to repeat tests, to death. Forty practising internists were asked to assess the probability that each event would occur. The expected cost of laboratory error was then calculated as the sum of the costs of each event multiplied by the probability of the event occurring.

2.3. *Explicit valuation*

Hospital costs were obtained from the price schedules of three hospitals. Physician costs were obtained from the sample of 40 internists who assessed the outcome probabilities.

3. Allowance for differential timing and uncertainty

Discounting would have been relevant to long-term consequences such as death and disability, but it was not clear whether these had been included.

Results were only calculated on the basis of the average responses given and no information was given on the distribution of the probability estimates.

4. Results and conclusions

Examples of the results obtained were given for both high and low error values. Except where the only consequence was repeat tests, low error values were more costly than high error values. For example, the probable costs of all consequences of high errors for prothrombin time and blood urea nitrogen were US$3944 and US$1056 respectively, whereas the corresponding costs for low values were US$11 170 and US$9007. Generally, errors on prothrombin time had the highest probable cost and cholesterol the lowest. The author concluded that this approach could be used to select those tests which should have highest priority for performance assessment as well as providing information for cost–benefit studies of improvement programmes.

5. General comments

The study illustrates a methodology for costing laboratory error and the uses to which it can be put. The basis of the study is expert judgment and it should perhaps be confirmed by collecting prospective information on observed consequences of laboratory errors.

3 Chapalain, M. T.(1978). Perinatality: French cost–benefit studies and decisions on handicap and prevention. In *Major mental handicap: methods and costs of prevention.* CIBA Foundation Symposium.

1. Study design

1.1. *Study question*

What is the most efficient way of using resources to reduce perinatal mortality and morbidity?(c)

Will a given package of measures achieve a given target (in reduced perinatal mortality, and at what cost?(a)

1.2. *Alternatives appraised*

Training of obstetricians and specialists in neonatology *versus* development of statistical information and research *versus* innoculation against German measles *versus* intensification of antenatal supervision *versus* improvement of childbirth conditions *versus* resuscitation in the labour room *versus* creation of intensive resuscitation units *versus* implicitly no programme.

1.3. *Comments*

This paper reported the use of the cost effectiveness analysis method within a broad, planning, programming, budgeting (PPBS) framework. Originally 60 possible interventions to reduce perinatal morbidity and mortality were proposed. These were reduced to 20 by further discussion and these then combined in the seven relatively independent sub-programmes analysed.

2. Assessment of costs and benefits

2.1. *Enumeration*

The costs considered for each programme included those to central government, local authorities and households. The effects considered were the number of deaths and the number of handicap cases averted. It was not clear whether the potential benefits of reduced mortality and morbidity (in terms of averted production losses and savings in health sector costs) were included.

36

2.2. Measurement

Few details were given of data sources in this paper. Comparisons were made of costs and effects over a 15-year period. The effects of the programmes were assessed independently. Since the programmes' impacts were not completely independent (e.g. improvement of risk factors may reduce the impact of curvative options), the data presented cannot be summed to give the total effect of the package of seven programmes.

2.3. Explicit valuation

It is likely that market prices were used to estimate costs. Although the effects of the programme were not valued in money terms, the two categories of effect were implicitly valued equally when combined in the index (cost)/(number of deaths *plus* number of handicap cases avoided).

3. Allowance for differential timing and uncertainty

It was not clear from the paper whether or not costs occurring in the future were discounted to present values.

No general sensitivity analysis was performed, although the problem of interdependence between the programmes was acknowledged by investigating two levels of effectiveness for the antenatal supervision programme (90 per cent and 50 per cent), *given* the concurrent implementation of the resuscitation programme.

4. Results and conclusions

Apart from the first two programmes listed, the seven sub-programmes could be ranked in terms of the cost of saving a life without after-effects. The most attractive programme (when considered in these terms) was resuscitation in the labour room, followed by the improvement (through increased supervision) of delivery and intensification of antenatal supervision.

It was concluded, furthermore, that the net costs of the total package of seven programmes would not be great (particularly if one considered the economic advantages of reducing mortality and handicap) and that in total it would achieve the government's chosen objective of reducing perinatal mortality from a projected 23/1000 births in 1980 to 18/1000 births.

Chapalain (1978)

5. General comments

Following the study reported, the programmes were implemented. The paper also discussed the follow-up results; the perinatal mortality rate did indeed fall, achieving the 1980 target by 1975. Although this cannot be linked unambiguously to the programmes, the fall was greater than in some other European countries over the same period.

While an important first step, the major methodological weakness of the study is the failure to assess the seven programmes in marginal terms. That is, what would be the additional contributions of each compared to the additional costs, given the presence of various combinations of the other six programmes.

4 Haberman, S. (1980). Heart transplants: putting a price on life. *Health and Social Service Journal* **90**, 877–9.

1. Study design

1.1. *Study question*

Can heart transplants be justified on cost–benefit grounds?(d)

1.2. *Alternatives appraised*

Immediate transplantation *versus* no transplantation with two years invalidity before death *versus* no transplantation with immediate death.

1.3. *Comments*

In the hypothetical examples considered, the patient was assumed to be 25 years old, male, married with two children and with gross earnings of £5000 per annum prior to the onset of illness. It might have made more sense to investigate two alternatives (to treat or not to treat), assuming varying probabilities (under the latter option) of the patient dying immediately or living for two years before dying.

2. Assessment of costs and benefits

2.1. *Enumeration*

The costs included comprised health care costs (the cost of the transplant operation itself, technician costs, inpatient and post-discharge care and nursing care for non-transplanted invalids), state benefits payable to dependents and the cost of lost patient earnings. Benefits for transplanted patients were calculated in terms of net earnings arising from a return to work.

No consideration was given to the benefits of either extending life *per se*, or to reducing any pain and suffering experienced by the patient.

Other than the income effects associated with changes in the patients' employment status, the effects of the tangible and intangible burden of illness on the patients' family were not considered.

2.2. *Measurement*

Survival and return to work probabilities were based on the experience of a series of 94 heart transplant patients at Stanford University. As only four years' survival experience was available from the Stanford data, it was assumed that the survivors of the transplant who return to work were 'normal' for their age as regards likely mortality experience. It was further assumed that those who survive and are incapable of returning to work have the likely mortality experience of an individual aged 20 years older.

Health care costs were based on cost estimates derived from the National Research Foundation and the Cambridgeshire Area Health Authority.

Expected benefits and costs were compared over a four and 10 year period.

2.3. *Explicit valuation*

Market values were used to estimate costs and benefits. Life years were valued in terms of the patient's net earnings (i.e. gross earnings less income tax and National Insurance contributions). This was supposed to represent the individual's contribution to the economy. However, this should be measured by the value of the output produced or gross employment cost (i.e. gross earnings *plus* the employer's National Insurance contribution).

3. Allowance for differential timing and uncertainty

The effect of including the range of variation in the US survival rates was demonstrated, as was the sensitivity of the results to the application of high and low operation costs. Costs and benefits were not discounted. This was an obvious omission, since costs and benefits occur over a period of time. (See the companion volume, Chapter 5.) The discussion of discounting was particularly confusing as it suggested, incorrectly, that the reason for discounting is to take account of inflation.

4. Results and conclusions

For all three alternatives the costs exceeded the benefits; the net cost of the transplant alternative was the lowest under all but the most pessimistic assumptions. Over a four year period, the net cost of the transplant alternative (using the higher hospital cost estimate and mid-survival estimate) was £28 060. Over the same period the net cost for non-transplant patients was estimated to be £38 880 for those surviving for two years and to be £29 740 for those who died immediately.

The corresponding net costs of the three alternatives over a 10 year period were estimated to be £36 800, £82 100, and £73 000 respectively. The increase in the gap between the transplant and non-transplant alternatives over the 10 year period was because of the greater loss of earnings and greater state benefits payable to dependants.

5. General comments

The study encountered problems of limited cost and epidemiological data. The two operation costs presented were drawn from different sources and differed by almost 100 per cent. Similarly, the net costs of the alternatives over a 10 year period necessitated extrapolation from the survival and return to work experience of a small sample of heart transplant patients, for whom only four year survival data were available.

The study took a very narrow approach to the value of human life (see the companion volume, Chapter 4 and Appendix 6). The study underestimated the benefits of transplantation in terms of the operation's ability to extend life. Conversely, the author's failure to discount future earnings (see the companion volume, Chapter 5) means that the costs of failure to treat were artificially high.

Similarly, by including both state benefits (transfer payments) given to the patient's family and the patient's loss of earnings due to illness, this may lead to an overestimate of the burden of illness attributable to a critical cardiac condition. (This will depend on whether the state benefits are more or less than the taxes netted out of earnings.) Also in including transfer payments, the study is closer to a *public sector financial appraisal* than an economic appraisal. That is, it attempts to investigate the impact of the three alternatives on the public purse, rather than looking at the costs and benefits to the community as a whole.

In seeking to justify heart transplants on economic grounds it is also necessary to compare the cost and benefits of transplants with those of other methods of saving life.

5 Hartunian, N. S., Smart, C. N., and Thompson, M. S. (1980). The incidence and economic costs of cancer, motor vehicle injuries, coronary heart disease and stroke: a comparative analysis. *American Journal of Public Health* **70**, 1249–60.

1. Study design

1.1. *Study question*

What are the economic costs of cancer, motor vehicle injuries (MVI), coronary heart disease (CHD) and stroke in the United States (estimated for 1975)?(b)

1.2. *Alternatives appraised*

Existing situation *versus* potential cost savings, should investments be made in the prevention of the disease in question.

1.3. *Comments*

The economic impact of disease and injury may be calculated by examining the costs associated with their *prevalence* in the reference year, or by assigning all present and future costs to the year of *incidence*. It is the latter approach which is adopted in this study. Alternative methods of calculating disease costs are discussed at length in Lindgen (1981) (summarised in this section).

2. Assessment of costs and benefits

2.1. *Enumeration*

The authors considered (under *direct* costs) expenditures on emergency services, hospitalization and outpatient treatment, drugs and medical supplies, medical equipment and appliances, home modifications, paramedical expenses, institutional and attendant care, and rehabilitation services. In addition, the administrative costs incurred by insurance companies and government agencies in funding illness and injury expenses were included. Under *indirect* costs the authors considered lost production attributable to patients' injuries and impairments (or premature death). The value of life *per se* and other intangibles were not considered, neither were costs falling on patients' families.

2.2. *Measurement*

Data were drawn from numerous studies reported in the literature.

The population was disaggregated by sex into eight age bands and by the principal conditions within the four health impairments considered. For example, in considering the incidence and costs of strokes, the authors distinguished between haemorrhagic and infarctive completed strokes and transient ischaemic attacks.

2.3. *Explicit valuation*

Market values were used to estimate costs in most instances. Two alternative methods were employed to value work in the home. One method was to value such work at the cost of employing domestic help; the other was to assign a value equal to the wage that persons working in their own homes would receive if they sought outside employment. Costs were expressed in 1975 prices.

3. Allowance for differential timing and uncertainty

Three discount rates (2, 6, and 10 per cent) were employed to convert future costs to equivalent present values. The effects of assuming three alternative real productivity growth rates of 0, 1, and 2 per cent per annum were examined in estimating the likely value of lost production resultant from the impairment. The effects of any likely errors in the epidemiological and cost data were also examined.

4. Results and conclusions

Of the four health impairments considered, MVI had the highest incidence; the reported number of MVI was 4.27 million, more than five times that of any of the other impairments considered. Assuming an annual growth rate of 1 per cent, a 6 per cent discount rate and valuing household labour by reference to domestic wage rates, the total costs of all cases were as follows: cancers, US$23 148 million; CHD, US$13 716 million; MVI, US$14 435 million; strokes, US$6456 million. These results were also presented by age and sex for the major constituent diagnostic sub-groups associated with the health impairments considered.

For all health impairments considered, indirect costs exceeded direct costs. The difference was greatest in the case of CHD, where the total indirect cost of US$11 225 million was 4.5 times the direct cost. It was least for strokes, where the indirect costs were only 1.7 times greater than the direct costs. This smaller difference arose because stroke affected relatively fewer productive years than did the other impairments considered.

Hartunian *et al.* (1980)

To estimate the impact of adopting the incidence approach to disease costing rather than the prevalence approach, the authors compared their results with those of Berk (Berk A. *et al.* (1978). The economic cost of illness, fiscal 1975. *Medical Care* **16**, 785–90). Using the prevalence approach, Berk estimated the annual cost of strokes to be US$6.84 billion, and that of cancer to be US$22.36 billion, whereas the incidence approach yielded costs of US$6.46 billion and US$23.15 billion respectively.

In examining the sensitivity of the results to the choice of discount rate, the estimated annual rate of productive growth and the method of valuing household labour, the authors concluded that within the range considered, the assumptions made could change the costs by a factor of two. The alternative results were reported in full in the paper.

5. General comments

The introduction to the study comprises a succinct summary of the major developments in the methodology of disease costing in the 1960s and 1970s and the extensive bibliography, from which the study draws, contains many useful references to disease costing methodology and studies.

A detailed description of the methods used in calculating the economic costs is given in Hartunian *et al.* (1980). *The incidence and costs of cancer, motor vehicle injuries, coronary heart disease and stroke: a comparative analysis.* Lexington Books, D. C. Heath and Co., Lexington, Ma.

6 Holtermann, S. and Burchell, A. (1981). The costs of alcohol misuse. *Government Economic Service Working Paper no. 37.* HMSO, London.

1. Study design

1.1. *Study question*
What is the cost to society of alcoholism and other forms of alcohol misuse?(b)

1.2. *Alternatives appraised*
Existing situation *versus* hypothetical eradication.

1.3. *Comments*
The authors point out that the real issue of interest is the marginal saving achieved by specific preventive measures. The disease costing approach used in the study is a preliminary step.

2. Assessment of costs and benefits

2.1. *Enumeration*
The authors attempted to cost all resource losses due to alcohol misuse. These included loss of output due to excess sickness, unemployment or early death, health and social services expenditures, damage due to accidents and law enforcement costs. Costs to families and intangibles were mentioned but not included. Some items had to be excluded because no data were available, for example social work costs and lost output due to impaired performance and absenteeism.

2.2 *Measurement*
The basic approach was to estimate the prevalence of alcoholism and problem drinking and the incidence of resource losses amongst these groups. Data were taken from surveys amongst alcoholics and supplemented by national statistics where available. The samples were often small and little information was available on problem drinking as opposed to alcoholism. The estimation procedures adopted were fully documented in the paper. The potential problem of double counting and the steps taken to avoid this were discussed.

2.3 *Explicit valuation*
All resource losses were costed at 1977 market prices. Lost output was valued at average gross employment costs, and lost household services were valued at average female earnings.

Holtermann and Burchell (1981)

3. Allowance for differential timing and uncertainty

Future earnings were discounted at 7 per cent to calculate the present value of the cost of premature deaths.

High and low estimates of costs were produced on difference prevalence assumptions. Other assumptions had to be made regarding the effects of alcohol misuse amongst non-alcoholics. Not all of these were tested for the sensitivity of the results, but some were. For example, it was calculated that if alcohol abuse in the general population caused one day of lost production per year for half the employed males, then an additional cost of £100 million per annum would be incurred.

4. Results and conclusions

The total costs of alcohol misuse were estimated to be between £428.2 million (low prevalence) and £650.1 million (high prevalence). However, various items had been omitted and the authors argued that these underestimate the true costs for each prevalence assumption.

The largest cost items were lost output due to sickness absence, £152.33 million to £253.82 million, lost output due to premature deaths, £138.2 million to £219.8 million and road traffic accidents due to alcohol misuse, £85.45 million. The authors pointed out that their estimates of the cost of alcohol misuse were equivalent to between one-half and three-quarters of the resource cost of road traffic accidents. Awareness of the size of the problem may lead to greater efforts to reduce the amount of alcohol misuse.

5. General comments

7 Johnson, W. G. and Heler, E. (1983). The costs of asbestos-associated disease and death. *Milbank Memorial Fund Quarterly* **61**, 2, 177–94.

1. Study design

1.1. *Study question*

What are the costs of asbestos-related diseases?(b)

1.2. *Alternatives appraised*

Existing situation *versus* (hypothetical) eradication.

1.3. *Comments*

Implicitly, this study is concerned with the potential benefits from reduced exposure to asbestos. However, a very narrow range of costs is considered, therefore the results must be considered as a minimum estimate.

2. Assessment of cost and benefits

2.1. *Enumeration*

The study was primarily concerned with private costs and concentrated on wage losses and losses of household production for workers who died before the end of their expected working life. Other elements of the full social cost were discussed but not included, mainly due to lack of data. The missing costs included medical care, losses in quality of life, additional deaths after the study period and family contact effects, i.e. deaths and disability amongst family members as a result of exposure to asbestos brought into the house.

2.2. *Measurement*

A cohort of 17 000 male insulation workers was followed-up for 10 years. Deaths due to asbestos-related diseases were noted and contact made with surviving widows to obtain information on social and economic consequences. Limited comparisons with agency records established the reliability of this information. Wage losses were based on age-specific expected working life (Bureau of Labour Statistics), adjusted for average hours worked to allow for unemployment. Wage loss is calculated both gross and net of taxes and workers' own consumption.

Johnson and Heler (1983)

2.3. *Explicit valuation*

Work loss was valued by reference to union wage agreements. Household production was valued by the national minimum wage rate.

3. Allowance for differential timing and uncertainty

There appeared to be some confusion in the presentation between discounting and conversion of costs from current prices to constant prices. The authors did both together, by applying nominal interest rates rather than real rates, but referred to this simply as discounting. As they were concerned with private costs, the authors argued that tax-adjusted market rates were appropriate and used trend values for US Treasury bonds.

The confusion over discounting led the authors to report both the actual nominal losses and the discounted present values. The results reported were clearly minimal and sensitivity analysis was not required.

4. Results and conclusions

Between 1967 and 1977 there were 995 deaths attributed to exposure to asbestos. Gross losses were calculated in 568 cases. The average loss per household from premature deaths was US$243 649. 468 workers experienced disability before death (average loss US$52 090). The overall average loss was US$252 331 and this was the minimum benefit to be obtained from saving the life of one worker. (All values given were discounted present values.)

The corresponding net losses were US$107 735 per death, US$41 022 per disability, and US$124 637 was the overall average net loss per household. These net losses were those experienced by the worker's family; by excluding taxes and the worker's consumption, the rest of society and the worker himself were excluded. These figures were used to examine the compensation paid to survivors. The median compensation rate was one-third of the loss to the survivor.

5. General comments

The authors attempt some extrapolation from their figures to the size of the overall problem of asbestos-related disease.

8 Lindgren, B. (1981). Costs of illness in Sweden 1964–1975. *Swedish Institute for Health Economics. Liber, Lund.*

1. Study design

1.1. Study question

What is the cost to society of all forms of ill health and of particular disease categories?(b)

1.2. Alternatives appraised

Existing situation *versus* (hypothetical) eradication.

1.3. Comments

The study also considered how costs have changed between 1964 and 1975. It aimed to measure the burden of disease, but did not consider treatment strategies.

2. Assessment of costs and benefits

2.1. Enumeration

All costs associated with prevention, detection, treatment, rehabilitation and long-term care due to disease and injuries were considered to be relevant. The author noted that, in theory, non-health care costs, should be included (for example, seatbelts, patients' travel costs). In practice, only health care costs (public and private) were considered.

Indirect costs were similarly restricted to the loss of output resulting from disease, disability and premature death, although the imputed value of non-marketed output was included. Intangible costs, such as pain and grief, were not considered.

2.2. Measurement

The indirect effects of short-term illness and permanent disability were measured by days of sickness benefit and numbers taking early retirement, respectively. Premature death was taken to be death before the official retirement age although it was acknowledged that people may work beyond this age.

The distribution of costs by disease category raised problems of joint costs for people with multiple diseases. Only the primary diagnosis was available and, therefore, joint costs could not be apportioned.

49

Lindgren (1981)

2.3. *Explicit valuation*

Health care services were valued at market prices, with aggregate costs taken from the Swedish national accounts. Average earned income (age and sex specific) was used to value production losses, both marketed and unmarketed. Unemployment was very low in Sweden and, therefore, no adjustment was thought necessary for this factor. (See Appendix 6 of the companion volume.)

3. Allowance for differential timing and uncertainty

Future production losses were discounted at 4 per cent per annum. Rates of 2.5 and 10 per cent were also applied to test the sensitivity of the results. The author discussed a number of assumptions made in carrying out the study and examined the effect of using cost per case rather than cost per patient day to allocate hospital costs between disease categories.

4. Results and conclusions

The total cost of illness in Sweden rose from Skr37 014 million in 1964 to Skr73 542 million in 1975 (1975 prices; 4 per cent discount rate). Health care costs accounted for 30 per cent of the total cost in 1975 (36 per cent, 1964), 52 per cent was due to morbidity (39 per cent, 1964) and 17 per cent of the cost arose from premature death (26 per cent, 1964).

The distribution of costs by disease category, for 1975, showed that diseases of the circulatory system, diseases of the musculoskeletal system and connective tissues, mental disorders, and accidents and injuries were the four largest cost categories and together accounted for over 50 per cent of total costs.

The substitution of cost per case for cost per patient day and the use of different discount rates had relatively little effect on the distribution of costs with only slight changes in relative positions. The use of a higher discount rate did, of course, reduce the absolute size of the production losses and vice versa.

5. General comments

The introduction to the study provides a review of other literature in this field and considers the alternative measurement approaches that can be adopted and the uses that can be made of cost of illness studies. The author supplements the results with an illuminating discussion of some of the underlying causes of changes in costs, for example, demographic trends and relative price effects.

9 Mills, E. and Thompson, M. (1978). The economic costs of stroke in Massachusetts. *New England Journal of Medicine* **229**, 415–18.

1. Study design

1.1. *Study question*

What are the lifetime direct and indirect costs incurred by patients in Massachusetts experiencing a first stroke in 1975?(b)

1.2. *Alternatives appraised*

Existing situation *versus* hypothetical prevention of stroke.

1.3. *Comments*

Information generated in the study was also used to compare the likely burden of disease associated with strokes, alcohol abuse, and smoking.

2. Assessment of cost and benefits

2.1. *Enumeration*

The direct costs considered comprised health care costs incurred inside and outside the hospital. Indirect costs of stroke comprised forgone earnings, calculated as the difference between the expected earnings of victims of strokes and those of the stroke-free population. The authors acknowledged that 'suffering and the effects of lowered life quality due to stroke are reflected only so far as they have economic consequences'.

2.2. *Measurement*

Statistics relating to incidence, recurrence rates and patients' post-stroke vocational status were drawn from various sources. In computing post-stroke earnings, the authors differentiated by type of stroke, i.e. cerebral infarction or haemorrhage.

Costs of in-patient treatment were based on patient charges in Massachusetts in 1975 and are likely to reflect average rather than marginal costs. Direct costs incurred by patients following hospital treatment (e.g. nursing home care, paramedical services and medical equipment, and appliances) were based on those reported in a previous paper, adjusted to 1975 price levels. (Elmet, H. E. Jr. *et al.* (1973). *Estimated health benefits and costs of post-onset care for stroke.* Analytic Services, Falls Church, Virginia.)

Mills and Thompson (1978)

2.3. *Explicit valuation*

Hospital charges were used to value the costs of in-patient care. The method of valuing other components of the direct costs of care was not stated, but presumably market prices were used. In calculating the subsequent lifetime earnings for the stroke and stroke-free population, a value was imputed for housewives' work, but the method used was not given.

3. Allowance for differential timing and uncertainty

The estimated lifetime costs of stroke were discounted at 10 per cent.

The authors point out that trends in stroke incidence and cardiovascular mortality were such that the estimates of stroke incidence used in the study might be rather high. No sensitivity analysis was conducted to show the likely effects of alternative stroke incidence rates on the results presented.

4. Results and conclusions

The present value of lifetime costs incurred by all patients in Massachusetts experiencing a first stroke in 1975 was calculated to be US$377.7 million, with direct costs accounting for US$204.6 million and indirect costs US$173.2 million.

Seventy-six per cent of all stroke costs (direct and indirect) were expected to derive from strokes occurring in the population aged over 55 years. Strokes in male victims as a group were expected to result in greater indirect than direct costs, reflecting reliance on lost earnings as the measure of indirect cost.

Survivors of stroke (at least six months after the incident) were expected to have average earnings equal to 49 per cent of those of comparable stroke-free persons. Patients suffering haemorrhage were expected to achieve lower future lifetime earnings than were those suffering cerebral infarction.

The direct costs of stroke in Massachusetts were shown to be 50.4 per cent of state health expenditures for alcohol abuse.

Nationwide, direct costs for stroke care would amount to 42.1 per cent of the health care costs due to alcohol abuse. However, when other costs of alcohol abuse (such as road traffic accidents and crime) were taken into account, the direct costs of stroke would only be 21.2 per cent of those resulting from alcohol abuse. National indirect costs of stroke were estimated to be 37.1 per cent of those for alcohol abuse in 1971.

Again on a national level, but in 1976 prices, the direct costs of stroke were estimated to be 72.3 per cent of those for smoking; indirect costs were 49.7 per cent of those for smoking.

5. General comments

The measure of indirect costs used here is a very limited one, especially given the large incidence of stroke amongst the elderly, who may not be in paid employment. For this group, lost earnings resulting from stroke would be expected to constitute a very small proportion of the total indirect costs borne by stroke victims and their families. Information on the comparative burden of diseases, such as that presented in the study, is useful to policy makers as one important determinant of resource allocation among competing client groups. However, there are other important factors in deciding upon such priorities.

10 Muller, A. (1980). Evaluation of the costs and benefits of motorcycle helmet laws. *American Journal of Public Health* **70**(6), 586–92.

1. Study design

1.1. *Study question*

Is it worthwhile making the wearing of motorcycle helmets compulsory?(d)

1.2. *Alternatives appraised*

Compulsory use *versus* voluntary use of motorcycle helmets.

1.3. *Comments*

The presentation of the study was slightly confusing, in that the costs and benefits of motorcycle helmet use were addressed in some detail before the question of the legal requirement to wear helmets was dealt with. Alternative methods of increasing helmet use were not considered in any detail although they were mentioned.

2. Assessment of costs and benefits

2.1. *Enumeration*

The only cost considered was that of the helmets themselves. Benefits were taken to be the averted costs of medical care and rehabilitation. The author recognized that this was a limited view of the costs and benefits. Additional costs such as the cost of compulsion and the costs of enforcing the law were discussed briefly. It was asserted that these would not be significant. Benefits such as life years saved and avoided pain and suffering were excluded mainly because of the difficulties of measurement or valuation.

2.2. *Measurement*

Two methods of measurement were used. First, the total costs and benefits of helmet use were calculated and a proportion of the net outcome was attributed to the increase in helmet use from legislation. Secondly, as an alternative approach, a comparison was made of the situation before and after the repeal of motorcycle legislation in two states.

The total number of helmets purchased was based on the average number of riders per motorcycle and the average replacement rate for used helmets. The fact that passenger helmets may be purchased but unused was ignored. The avoidance of injuries was measured by the difference in the expected pattern of facial and head injuries experienced by helmeted and non-helmeted motorcycle riders, based on the Overall Abbreviated Injury Scale for the two groups.

2.3. *Explicit valuation*

The cost of helmets was taken to be the average purchase cost. Medical care and rehabilitation costs for each injury class were taken from the US Department of Transport estimates for the costs of motor vehicle accidents.

3. Allowance for differential timing and uncertainty

Differential timing was not important because the future benefits (e.g. resulting from the life years saved) had been excluded.

A number of estimates were used, for example, in calculating the replacement rate of helmets and the accident rate. These should have been subjected to sensitivity analysis. The author discussed the possibility that voluntary helmet wearers have different attitudes to risk and hence have different accident rates, but the implications for the results were not tested.

4. Results and conclusions

Total annual consumer expenditure on helmets was between US$766 871 and US$910 194 per 100 000 motorcycles. Averted medical care costs were estimated to be US$2 094 050 per 100 000 motorcycles, implying a net benefit of at least US$1 183 856 per 100 000 motorcycles.

Helmet use was estimated to rise from 50 to 95 per cent when legislation was introduced. The net benefit attributable to legislation was therefore between US$597 231 and US$532 735 per 100 000 motorcycles. The comparison of the pre-repeal and post-repeal situation in two states provided an alternative estimate of the net additional cost after repeal, which was between US$644 789 and US$718 170.

5. General comments

The author argued that by ignoring intangible benefits, the net benefit would be underestimated.

Muller (1980)

If all the costs were included and the net benefit was still positive, then the additional effort of measuring and valuing intangibles would not be necessary. This is a perfectly sound approach, but there are some reservations about the comprehensiveness of the range of costs in this study.

1 Muller, C. F. and Caton, C. L. M. (1983) Economic costs of schizophrenia: A postdischarge study. *Medical Care* **11**(1), 92–104.

1. Study design

1.1. *Study question*

What are the costs of chronic schizophrenia after discharge from hospital?(a)

1.2. *Alternatives appraised*

Costs for schizophrenics *versus* (implicitly) non-schizophrenics.

1.3. *Comments*

The design of the study was not always clear. The objective appeared to be to identify the potential for making de-institutionalized care more cost-effective but this needed to be brought out more strongly. Cost data were collected for a sample group of discharged schizophrenics but the comparison with non-schizophrenics was made on a hypothetical basis. This gave rise to some confusion, for example, the authors intended to include general medical care costs but it is not clear that they would have identified the *difference* in these costs for schizophrenics *vis-à-vis* non-schizophrenics. (The figures were not included because of incomplete data.)

2. Assessment of costs and benefits

2.1. *Enumeration*

The direct costs associated with discharged schizophrenics were for community aftercare, ambulatory health care and rehospitalization. The indirect costs were wage costs and the value of lost homemaker services. No attempt was made to include diminished health status, *per se*.

2.2. *Measurement*

A variety of data sources were employed, including preadmission economic histories, quarterly questionnaires covering psychological, economic, and social adjustment and use of services and provider medical records. Self-reporting by patients was cross-checked against institutional records. Wage loss was measured by considering the hypothetical earnings of a group of non-schizophrenics with the same

Muller and Caton (1983)

demographic characteristics. In this way, the authors controlled for work loss among 'normal' earners.

2.3. *Explicit valuation*

'The definition of cost is based on market prices or accepted reimbursement schedules; the point of view is that of the agency or individual buying services and the measure of interest is what such buyers are obliged to pay.' The value of homemaker services was equated with average market wages.

3. Allowance for differential timing and uncertainty

The costs considered were for the first year after discharge only, therefore discounting was not required.

The only sensitivity analysis that was conducted was the reporting of upper and lower limits for production loss by including or excluding household services. The authors do not report the distribution of the results around the averages reported but do report costs for various classes of patient where the difference in cost was significant.

4. Result and conclusions

The average direct cost was US$7125 (US$2020 for community care and US$5105 for hospital care). Average wage loss amounted to US$5760. The average value of household services lost was US$833. The main factor contributing to direct cost was the high rate of rehospitalization. Fifty-nine per cent of patients were readmitted within the year.

Costs were significantly higher for men than women but this was primarily due to the higher loss of earnings. Patients living in single room occupancy dwellings had higher wage loss than other groups (but the figures were not adjusted for sex differences). The authors conclude that deinstitutionalized care could be made more cost effective by programmes designed to improve employment prospects and to reduce readmission to hospital. However, they present no real evidence to support this assertion.

5. General comments

The paper includes a review of earlier literature on the costs of mental illness.

2 Yelin, E. H., Fishbach, D. M., Meenan, R. F., and Epstein, W. V. (1979). Social problems, services and policy for persons with chronic disease: the case of rheumatoid arthritis. *Social Science and Medicine* **13c**, 13–20.

1. Study design

1.1. Study question

What are the social and economic implications of severe rheumatoid arthritis (RA)?(b)

1.2. Alternatives appraised

Existing situation *versus* (hypothetical) elimination of disease.

1.3. Comments

A sample of 50 patients with severe chronic RA (third stage) were interviewed, and information collected on tangible and intangible costs, supplemented by costs derived from outside sources. The sample is rather small and may be biased in that it was drawn from medical practices rather than the general community.

2. Assessment of cost and benefits

2.1. Enumeration

Only costs and changes associated with RA were considered. Costs of medical and social services utilized and income losses of wage earners and housewives were calculated. Categories of services utilized included number of visits to rheumatologists, all MDs, and non-MDs including social services, days spent in hospital, and number of laboratory tests and X-rays. The percentage of patients experiencing changes in work status, and suffering prolonged disability or other psychosocial changes was also calculated but not costed. All calculations were broken down into patient groups, according to individual and family income, pre-morbidity occupation and disease capacity and duration.

2.2. Measurement

The frequency of utilization of health and social services was calculated and averaged by patient category. Costs of MD visits, drugs, tests, and out-patient and in-patient visits by category of patients were calculated to arrive at total medical costs by patient category. Patient questionnaires, which included financial information, were supplemented by information from medical records and hospital bills. Percentage of individual's income loss and family income loss, and also actual income loss were calculated by patient category. All costs and losses were expressed as averages.

2.3. *Explicit valuation*

The cost of medical services, was determined by applying average unit costs to all categories. Costs of services not common to all the sample and of hospitalization were determined from bills received by patients. Income losses for wage earners were estimated by taking premorbidity earnings and dividing by the median earnings for US workers of the same sex and occupation to obtain a relative wage, which was then updated by reference to the median income for year under study (1976). This gave an estimate of expected income which was compared with actual income to calculate income loss. Income loss for housewives was calculated separately using a method which values housewives' activities at market rates during different stages of their lives.

3. Allowance for differential timing and uncertainty

All results were expressed as a mean, with a range of plus or minus one standard deviation, and by patient category. No other sensitivity analysis was carried out. Costs were considered for one year only and as no specific eradication measures were being examined, the issue of discounting did not arise.

4. Results and conclusions

All groups of patients were heavy users of medical services, but infrequent users of social services. There were no significant differences across patient groups. An average RA patient spent US$2323 on medical and social services, three times the national average. Because of their lower incomes, medical costs accounted for one-quarter of annual income for the average RA patient. Income losses averaged US$6830 per annum, three times the average medical costs. An RA patient lost, on average, 61 per cent of expected earnings. This was ameliorated to some extent by family earnings, in that family earnings were reduced by only 37 per cent on average. Income losses were significantly different across occupational groups. Finally, between one-half and three-quarters of RA patients experienced changes in work, family, and leisure status.

Overall, therefore, social costs for third stage RA exceeded medical costs, and the use of medical services exceeded those for social services.

5. General comments

The authors note the need for further work to be built on their findings. In particular, the question of whether or not increased use of social and/ or medical services would reduce the social costs of RA needs to be addressed. The authors correctly identify that the policy issue is to select the appropriate mix of services but they also recognize the difficulties that exist in implementing policy even if the best service mix can be specified.

SECTION 2: Alternatives in prevention

Introduction

The studies summarized in this section are of a wide range of preventive options. Six of the studies concern *primary prevention* programmes such as vaccination against particular diseases (e.g. pneumococcal pneumonia, pertussis, hepatitis B and, in developing countries, measles, diphtheria, tetanus, and tuberculosis) or dietary change. Another seven studies contain evaluations of *secondary prevention* measures such as screening for high cholesterol levels, hypertension, phenylketonuria in newborns, screening the unborn for spina bifida, Tay–Sachs disease, and congenital toxoplasmosis. A further seven studies concern preventive measures that can be undertaken as part of medical therapy or by medical means. These include electronic fetal monitoring, ECG monitoring in the operating room, dental hygiene practices, drug prophylaxis to prevent infection following hysterectomy, prophylaxis for pulmonary embolism in high-risk surgical patterns, and leucocyte transfusion in the treatment of acute leukemia. Some of the studies in this last group could easily have been included in the diagnosis or therapy sections of this volume and the methodological points discussed there also apply. In addition, a number of the preventive options summarized here include, as part of the strategy, treatment for the cases found.

However, the common thread in all the studies considered is that they involve interventions that are made *earlier* than those that would be required purely on the basis of curative considerations. That is, they result from a belief that earlier intervention is more beneficial, in that it averts unnecessary mortality or morbidity, or less costly, in that treatment costs are averted. (Some preventive measures would claim to be both more beneficial *and* less costly.)

Depending on the focus of the studies, they consider various combinations of the costs of mounting the prevention programme plus any associated treatment costs, the improvements in health obtained and the value of those improvements in savings in health service resources, in productivity gains or in better quality of life.

Particular methodological problems in this area

One of the major problems in evaluating preventive measures is in mounting controlled studies to assess the effectiveness of interventions. This is particularly true of the primary prevention measures applied to whole populations, such as vaccination schemes or mass media campaigns to change diet or lifestyle. Whereas on some occasions it may be possible to obtain a control group from another location, in most cases evidence on the effectiveness of primary prevention measures is either obtained from 'before and after' studies, or by extrapolating from evidence obtained under more ideal circumstances, e.g. experiments to establish the efficacy of a vaccine. Therefore, it is common for evaluations in this field to employ extensive sensitivity analysis. (See Chapter 2.) Of course, there is no compelling reason why the effectiveness of secondary prevention measures, or those delivered as part of medical therapy, should not be assessed by controlled trials. Indeed a number of studies summarized in this section do employ such a methodology. (For more comments on controlled trials see Chapter 2 and the introduction to Section 4.)

Other methodological difficulties stem from the fact that many of the benefits of preventive measures accrue in the future. Three points are worth noting in particular. First, since the potential improvements in health are often far into the future, some studies consider intermediate output measures such as 'number of cases averted' or 'number of individuals at risk identified'. While this is partially acceptable, supplementary evidence on the link between changes in the intermediate measures and long-term outcome is required.

Secondly, it may be that while a preventive intervention may control, say, coronary risk factors, other events, such as a more general change in tastes in the future, may have controlled these in any case. In their evaluation of cholesterol control programmes for children in the United States, Berwick et al. (1981) noted that there had been significant declines in cardiovascular disease in recent years and that a continuing decline would have an impact on the performance of preventive programmes. Technological advances in medicine would also have an impact. Of course, the fact that in most analyses costs and benefits occurring in the future are discounted to present values already gives them a lower weight in the analysis. In addition, if there were evidence on the probability of changes in tastes or technology this could also be incorporated through sensitivity analysis or by an expected value calculation (Drummond 1980). However, since the decision to implement the preventive programme has to be made at a given point in time, it can only reflect knowledge at that particular point in time. Nevertheless, under highly uncertain conditions it would make sense to select a strategy that would be more robust if such changes did occur (Rosenhead 1978).

Thirdly, the fact that many of the benefits of preventive programmes occur in the future means that the choice of discount rate has a greater impact on study results. This was also noted by Berwick *et al.* (1981) in that the choice of discount rate affected the cost per year of life gained from cholesterol screening and hence the apparent 'worth' of the screening programmes compared with those in other fields. It is much less common for the choice of discount rate to affect the ranking of programmes within a given field, however. It did not do so in the case cited. As was mentioned in Chapter 2, the most sensible approach is to use a range of discount rates and to highlight the impact of the choice of rate in the discussion of study results.

Before leaving the discussion of the long-term nature of many of the benefits of prevention programmes, it is worth noting that Cohen and Mooney (1984) argue that we need not necessarily wait until the future for all the benefits. They argue that many programmes confer benefits in 'utility-in-anticipation'. That is, one does not have to suffer an attack of a potentially preventable disease in order to derive benefits from being protected from it; similarly, in driving along a divided road one might take comfort from the existence of crash barriers even though the probability of being involved in an accident is small and the accident, if it were to occur, may happen far into the future. The whole area of the value of reassurance and the disutility of increased anxiety resulting from some prevention programmes is one where estimation procedures are not well advanced.

The final group of methodological problems to be discussed here concern 'spillovers', in other health care costs or in other sectors of the economy, resulting from prevention programmes. For example, if a programme for screening and treating hypertensives results in them living longer, there may be increased health care costs in the future owing to them suffering from cancer and arthritis. (See Stason and Weinstein 1977.) Also, if prevention programmes aimed at cardiovascular disease are successful in changing diet and lifestyle, there may be an economic impact on the agricultural industry.

Economists are divided on the issue of whether increased health care costs in later years should be included in the analysis. However, it should be noted that in any case the inclusion or exclusion of such costs is unlikely to have much impact on the analysis since they will be heavily discounted. Nevertheless, in principle they should be included if the analysis also considers the benefits, in improved quality of life, that such treatments confer, e.g. hip replacements. It would be wrong, however, to dismiss preventive strategies because they merely postpone, rather than avert, health care costs. Such a view ignores the substantial benefits of increased quantity and quality of life.

On the second issue, of spillovers into other sectors of the economy, it may be sufficient merely to note these if they are small. However, it might be argued that the effects of some programmes on the supply and demand of goods and services, and on relative prices, are so extensive that the *partial equilibrium* approach embodied in economic evaluation is inappropriate. The normal assumption in economic evaluation is that projects are small in the sense that they will not alter the constellation of relative prices. Where this assumption does not hold one must go over to a *general equilibrium* approach, where the economy of the country is modelled so as to estimate the broader effects of the programme on the economy as a whole. For example, in an evaluation of malaria control programmes, Barlow (1967) argued that the effects on mortality and morbidity experience, fertility and agriculture were so far-reaching that a *general equilibrium* approach was required. However, in practice, few analysts adopt this approach, although it should be incumbent on those undertaking economic evaluations of preventive strategies having substantial spillovers to note these in the discussion of study results.

Current state of the art

First, it is interesting to note that many of the studies reported here compare the preventive programme with the implicit alternative of 'doing nothing'. Few studies investigate alternative strategies for prevention of the given disease and it is therefore not clear whether the most cost-effective strategy has been selected for evaluation. Notable exceptions in this regard are the studies by Berwick *et al.* (1981), Hull *et al.* (1982), Mulley *et al.* (1982), and Stason and Weinstein (1977), all of which consider a wide range of options. Another problem which arises in studies implicitly comparing the preventive programme with 'doing nothing' is that the latter option is often inadequately described. It is therefore sometimes difficult to establish the baseline against which the preventive programme is being assessed and whether 'doing nothing' could be organized more efficiently, or is likely to change through time as a result of technological advances.

Estimation of the costs of preventive programmes is relatively straightforward, although many of the studies summarized here fail to give sufficient details of costing method. In particular, it is often hard to ascertain whether capital costs have been included. (This deficiency is also present in many of the studies summarized in other sections.) In costing a prevention programme, the costs of treating side-effects (e.g. of vaccine) should be included. In addition, in the case of screening programmes, the cost of treating the cases detected should clearly be considered as a integral part of the prevention programme. Most of the studies summarized here include such costs in so far as they are incurred by the health sector, but few consider costs falling on patients (e.g. time costs in obtaining a screen and in receiving treatment). However, in many cases inclusion of patients' costs would be unlikely to change study results, since such costs may be also large in the curative option with which the prevention programme is being compared. Nevertheless, authors ought to make it clear that this is the assumption being made in their study.

Estimation of benefits is sometimes in physical units such as years of life gained (e.g. Berwick *et al.* 1981), but also in quality-adjusted life years (e.g. Stason and Weinstein, 1977; Willems *et al.* 1980). Some studies use the human capital approach, based on discounted future earnings, to estimate the benefits of prevention programmes. The reader should consult Chapter 2 and Drummond (1980) for further discussion of these approaches.

Finally, given the uncertainties surrounding the costs and benefits of prevention programmes, it is pleasing to note that many of the studies employ extensive sensitivity analysis. Particularly good examples are to be found in the studies by Berwick *et al.* (1981), Henderson (1982), Koplan *et al.* (1979), Mulley *et al.* (1982), and Stason and Weinstein (1977).

Contribution to decision making

There is some evidence that economic evaluations in the field of prevention have helped to formulate policies in this area. This is particularly true of some of the appraisals of vaccination programmes in the United States, (e.g. pneumococcal pneumonia and pertussis). This is partly because, on the basis of the studies carried out, it is possible to construct a strong argument that from the government's point of view the costs of mounting the programme are far outweighed by the resource savings. Some of the studies discuss refinements in policy based on the marginal costs and benefits of vaccinating different groups in the community. For example, in their study of hepatitis B, Mulley *et al.* (1982) argued that vaccination of homosexual men and surgical residents would result in savings in medical care costs, but that neither screening nor vaccination would be the lowest cost strategy for the general population. (However, they pointed out that the costs and savings were dependent upon the annual attack rate and that broader indications for the use of hepatitis B vaccine would be well justified if production losses were considered important, notwithstanding the benefits of averting morbidity and mortality *per se*.)

Of course, even in cases where it can be demonstrated that from the government's viewpoint the savings outweigh the costs, the decision to implement, say, a vaccination programme is not straightforward. First, few of the studies discuss whether the savings they identify will, in fact, be realized, by cutting back on beds or health service manpower, or whether the 'savings' should be interpreted as the *potential* benefits that could be obtained from redeployment of the freed resources. (See Chapter 2.) Koplan (1985) cites the case of fluoridation in the United States where steps were taken to reduce the supply of dental graduates by cutting back on dental school intake. On a more micro level, Shapiro *et al.* (1983) point out that while their study showed that the use of prophylactic cefazolin for 8–24 hours in patients undergoing hysterectomy reduced costs, routine practice (of four to five days) would not result in similar savings.

Secondly, it should be recognized that whilst it can be fairly readily established that many vaccination programmes generate benefits to the community in excess of the costs, a minority of individuals are put at risk (e.g. from pertussis vaccine). Therefore in deciding on how strongly to urge the general public to participate in vaccination programmes, the policy maker needs to consider this ethical issue. Whilst not prejudging the ethical question, economists would argue that where there are significant distributional impacts of programmes, compensation should be more frequently considered. That is, if there are net benefits to the community as a whole, there should be some redistribution of the overall gains to those who lose out.

This raises a further point that is relevant when considering the policy relevance of economic evaluations of preventive strategies. Namely, equity considerations are likely to be particularly important when evaluating options in the prevention field. Indeed, the desire to help disadvantaged groups in society may be a major motivating factor in the development of screening and vaccination programmes. Therefore, economic evaluations would probably benefit from some explicit consideration of distributive issues, e.g. who benefits from the prevention programme: is it effective in reaching the poor or otherwise disadvantaged individuals? In one study summarized in the earlier volume (Drummond 1981) this issue was discussed when two methods of screening were compared (Rich *et al.* 1976). It is also raised, in the context of services for rural and urban communities, by Ponnighaus (1980). However, these two studies are the exception rather than the rule.

Furthermore, it might be argued that some preventive programmes differ from health care programmes more generally, in that as well as being directed at improving health status they are also instruments of social change. This is especially true of those programmes that attempt to change lifestyle, by changing information flows to the general public or by changing incentive structures. It is probably also true of those programmes that restrict individual freedom in the interests of the common good.

References

Barlow, R. (1967). The economic effects of malaria eradication. *American Economic Review* **57**, 130–47.

Cohen, D. R. and Mooney, G. H. (1984). Prevention goods and hazard goods: a taxonomy. *Scottish Journal of Political Economy* **31**(1), 92–9.

Drummond, M. F. (1980). *Principles of economic appraisal in health care.* Oxford Medical Publications, Oxford.

—— (1981). *Studies in economic appraisal in health care.* Oxford Medical Publications, Oxford.

Koplan, J. P. (1985). Benefits, risks and costs of immunization programmes (discussion). In *The value of preventive medicine.* Ciba Foundation Symposium No. 110 (Chairman: A. G. Shaper). Pitman, London.

Rich, G., Glass, N. J., and Selkon, J. B. (1976). Cost-effectiveness of two methods of screening for asymptomatic bacteriuria. *British Journal of Preventive and Social Medicine* **30**, 54–9.

Rosenhead, J. (1978). An education in robustness. *Journal of the Operational Research Society* **29**(2), 105–11.

3 Banta, H. D. and Thacker, S. B. (1979). Assessing the costs and benefits of electronic fetal monitoring. *Obstetrical and Gynaecological Survey* **34**(8), 627–42.

1. Study design

1.1. *Study question*

What are the costs and benefits of providing electronic fetal monitoring (EFM)? (d)

1.2. *Alternatives appraised*

Electronic fetal monitoring *versus* (implicitly) doing nothing.

1.3. *Comments*

Total costs of providing the service in the United States were estimated, and benefits were assessed by reference to the existing literature.

2. Assessment of costs and benefits

2.1. *Enumeration*

The financial cost of providing EFM in the United States was estimated. Only the direct and indirect costs associated with EFM were included. Direct costs included medical care costs of delivery; costs of treatment for maternal and infant complications such as Caesarian section; and expenditure on health services, such as costs of care for mentally retarded individuals. Indirect costs included the value of lost production resulting from changes in maternal and infant mortality and morbidity.

2.2. *Measurement*

The study was not linked to a clinical trial. By reference to the literature, the number of deliveries using EFM was estimated, plus the number of associated Caesarian sections, and incidence of morbidity and mortality. The average cost of each case of Caesarian section, and each case of neonatal and maternal morbidity was calculated, and hence the total costs associated with EFM were calculated. The indirect costs associated with maternal and neonatal deaths were calculated by taking the present value of lifetime earnings for women (aged 30–34), and for males and females at age of one year.

2.3. *Explicit valuation*

Relevant costs were derived from current literature; no original costs were calculated. Monetary values were not placed on the benefits.

3. Allowance for differential timing and uncertainty

A discount rate of 10 per cent was used in calculating the net present value of lifetime earnings. Where ranges of costs were given, only the average was used in calculation.

4. Results and conclusions

The estimated financial cost of EFM to the USA (at 1977–78 prices) was US$411 million. Doubt was cast on the efficacy of EFM as a widely used monitoring tool. The US$411 million spent on EFM was compared to the US$80 million spent on all childhood immunization programmes.

5. General comments

This was mainly a clinical assessment of a monitoring procedure with reference to a large body of literature. No alternatives to EFM were clinically or economically assessed and the benefits of EFM were not explicitly calculated. Therefore, the conclusions drawn are very limited. Does US$411 million spent on EFM represent 'value for money'? This study does not provide an answer. (See also Thompson and Cohen 1981, summarized in this section.)

Barnum, H. N., Tarantola, D., and Setiady, I. F. (1980). Cost-effectiveness of an immunization programme in Indonesia. *Bulletin WHO* **58**(3), 499–503.

1. Study design

1.1. *Study question*

Given the levels of incidence in Indonesia, is it cheaper to treat certain infections or to prevent infection through an immunization programme? (c) (See 1.3 below.)

1.2. *Alternatives appraised*

Expanded programme of immunization (involving an extension of the programme to cover a larger proportion of the population and to include immunization for diphtheria, pertussis, and tetanus) *versus* existing immunization programme.

1.3. *Comments*

Although the study question is classified as (c), the study, in presenting results in terms of cost per death averted, is hinting at the question: 'Is treatment worthwhile *per se*?'

2. Assessment of costs and benefits

2.1. *Enumeration*

The study considered only the direct (health service) costs of the immunization programme and the 'curative' alternative. The immunization programme costs included both capital and operating components, whereas the costs of the curative option omitted capital costs in hospitalization. Health effects other than deaths averted were not considered (see 5 below), neither were production losses.

2.2. *Measurement*

Assessments of effectiveness of the programme (in cases and deaths averted) were made by extrapolation from current knowledge and by technical judgment. There was no prospective study. In the view of the authors, conservative estimates of effectiveness were made.

2.3. *Explicit valuation*

Costs were valued using market prices. For example, hospital costs were calculated by estimating length of stay and then multiplying by a daily charge.

Barnum *et al.* (1980)

3. Allowance for differential timing and uncertainty

No allowance was made for differential timing of costs and effects. For example, a straight line depreciation method was used to estimate annual capital costs, rather than an annuitization method. (See Chapter 2.)

A limited sensitivity analysis was performed, varying costs upwards by 20 per cent and benefits downwards by 50 per cent.

4. Results and conclusions

It was concluded that an expanded programme (including immunization for diphtheria, pertussis, tetanus, and tuberculosis) would be highly cost effective in comparison with treatment, i.e. a cost per death prevented of US$130. Separate analysis of the DPT-tetanus toxoid and BCG components of the programme showed that 'although the BCG programme may not be justifiable when operated independently, its inclusion in a *joint* immunization programme is strongly justifiable on economic grounds.' (See 5 below.)

5. General comments

A number of the methodological issues arising in the estimation of the costs and benefits of immunization programmes are discussed in Creese, A. L. and Henderson, R. H. (1980). Cost–benefit analysis and immunization programmes in developing countries. *Bulletin WHO* **58**(3), 491–7.

In common with a number of studies in this area, the outcome measure chosen is 'deaths averted' rather than 'years of life' or 'quality-adjusted life-years' gained. This is particularly important since vaccination against a few diseases may not increase life expectancy very greatly in a population with high risk of other causes of death.

The study makes the point that, once the system for refrigerated delivery of vaccine (the 'cold chain') and the system of vaccine administration are in place, the *incremental* costs of adding extra programmes are small. However, the presentation of results in the paper is unclear in this respect.

15 Berwick, D. M., Cretin, S., and Keeler, E. (1981). Cholesterol, children, and heart disease: an analysis of alternatives. *Pediatrics* **68**(5), 721–30.

1. Study design

1.1. *Study question*

What is the most cost-effective method of increasing life expectancy by controlling cholesterol levels in children? (c) and (d) (See 1.3 below.)

1.2. *Alternatives appraised*

Universal screening plus dietary counselling in 10-year-olds *versus* targeted screening with dietary counselling *versus* population-wide intervention through mass media *versus* doing nothing.

1.3 *Comments*

The authors point out that the beneficial effects of lowering cholesterol levels through diet were not proven at the time of the study. However, they argue that 'the burden of cardiovascular disease in the United States is so great that it may nonetheless be reasonable to initiate measures now, rather than to wait for proof'. The 'doing nothing' alternative was included as a reference programme to which others were compared. It was assumed to have zero costs and zero benefits. In addition, variants of the four main options were discussed.

2. Assessment of costs and benefits

2.1. *Enumeration*

The costs considered were those public sector resources that would be used in the hypothetical programmes being evaluated. These included the cost of screening and of treating those whose screen is positive (i.e. counselling them to change their diets and following up on their cholesterol levels). For population-wide interventions the costs included those of media or educational efforts. The costs to clients of participating in the programmes were not considered. The effects considered were the years of life saved by the interventions.

2.2. *Measurement*

The study was not linked to a prospective controlled clinical evaluation.
The effects of reductions in cholesterol level on the risk of cardiovascu-

lar disease were estimated from the Framingham Study of Heart Disease. These age-specific risks were combined with competing risks of death from other causes and with estimates of fatal and non fatal heart disease rates to calculate 'years of life remaining' and 'years of heart-disease-free life remaining' as a function of cholesterol level. The effects of dietary changes on cholesterol levels were also obtained from the literature.

2.3. *Explicit valuation*

The costs were estimated using the market prices of the hypothetical programmes (in 1975, US dollars). No attempt was made to estimate the value of years of life gained or to assess the quality of that life. This did not affect the choice between the three programmes examined as the quality of the years of life saved was the same for each. It did mean, however, that the analysis could not directly answer the question: 'Is any intervention in this field worthwhile?' The authors began to address this point by comparing the results of this study, in cost per year of life saved, with the results obtained in the evaluations of other health care programmes.

3. Allowance for differential timing and uncertainty

Costs and effects were discounted at 5 per cent in the 'base-case' estimate. The authors also presented the effects in undiscounted form, arguing that 'discounting of health benefits is still controversial' and that 'the correct rate of discounting is also in dispute'. However, they pointed out that their belief was that the discounted numbers were more appropriate when comparing different programmes to each other. (We agree with this assessment. See Chapter 2.)

An extensive sensitivity analysis was performed. The impact of variation in the following factors was explored: the degree to which risk depends on cholesterol level, the stability of cholesterol level, compliance fraction, the change in cholesterol with diet, the cost of screening, the cost of the dietary intervention, and the discount rate. The likely range for many of these factors was taken from the available literature. Finally, the authors listed a series of additional factors, not fully considered in the analysis, that should be brought to bear when interpreting study results.

4. Results and conclusions

The most cost-effective programme was the community-wide intervention through mass media; this cost around US$2800 per year of life saved at a discount rate of 5 per cent. By contrast, a cholesterol screening programme for all 10-year-old children would cost about US$10 000 per year of life saved. Targeted screening of high-risk children would improve the efficiency of the screening option by about 25 per cent, but would benefit only one sixth as many people. The cost per year of life saved was most affected by the rate of discount and the dollar cost of changing behaviour, but was insensitive to the cost of screening.

5. General comments

This study is discussed in more detail in Berwick, D. M., Cretin, S., and Keeler, E. (1980). *Cholesterol, children and heart disease: an analysis of alternatives.* Oxford University Press, New York.

The main strengths of this study are the use of sensitivity analysis, the clear exposition of the policy options and the identification of the main caveats on the analysis. The estimation of the cost of the preferred option—the mass media campaign—could be discussed in more detail. However, the authors point to the fact that these could approach double the amount observed in previous mass media compaigns without changing the ranking of the options. Another potential deficiency, identified by the authors, is the failure to estimate the impact of a continuation of the observed fall in cardiovascular morbidity in the United States.

Finally it is worth pointing out that this study is one where programme performance is crucially dependent on the discount rate.

16 Congress of the USA (1981) Office of Technology Assessment. *The implications of cost-effectiveness analysis of medical technology. Case study 5: Periodontal disease: assessing the effectiveness and costs of the Keyes technique.* Washington, DC.

1. Study design

1.1. *Study question*

Is the use of the Keyes technique worthwhile in preventing the development of periodontal disease and the need for restorative surgery?(d)

1.2. *Alternative appraised*

Use of the Keyes technique *versus* (implicitly) 'doing nothing'. (See 1.3 below.)

1.3. *Comments*

Specific treatments were discussed in the paper but no comparison was actually carried out. The study reported only the difference in patient condition before and after treatment with the Keyes technique. Various alternative methods of patient management are available. The Keyes technique is a regimen involving several treatment and oral hygiene practices. As no controlled comparison was carried out, the study could not identify the extent to which the Keyes technique contributed to the prevention of deterioration that would have otherwise occurred.

2. Assessment of cost and benefits

2.1. *Enumeration*

The costs of dental treatment consisted of dentist's and hygienist's time, plus special equipment, e.g. phase-contrast microscope. The dental practice was considered to be already in operation and fixed costs were therefore excluded. Purchases by the patient for treatment purposes were included, but travel and time costs were not.

2.2. *Measurement*

Data were collected from 18 practices using the Keyes technique. A sample of about 10 patients from each practice was taken. The extent of periodontal disease before and after treatment was measured by five indicators and effectiveness was also measured by the ability of patients to control plaque after treatment.

2.3. *Explicit valuation*

The cost of dentist's and hygienist's time was valued on the basis of yearly income and hours worked. Equipment and other purchases were valued at market prices.

3. Allowance for differential timing and uncertainty

Capital equipment specifically required by dentists for the Keyes technique was depreciated over a 10 year period. The method used was not given, although the authors pointed out that the capital equipment cost had very little impact on their estimates. (See Chapter 2 for details of the annuitization procedure for capital equipment costs.)

4. Results and conclusions

The results showed a reduction in the percentage of patients with indications of periodontal disease and an improvement in plaque control. The estimated average variable cost of a dental visit was between US$17.87 and US$13.72 and the treatment consisted of six visits with subsequent maintenance visits. Additional purchases by the patient cost between US$30 and US$60. The authors report that only between 0 and 5 per cent of patients from the study practices were referred for periodontal surgery but no comparative referral rates for non-treated groups were given.

5. General comments

The study exemplifies the difficulty of appraising clinical practices when the epidemiological base is unsatisfactory. The authors stress the lack of evidence about effectiveness of alternatives and recommend a randomized trial. Their views are challenged in an independent commentary on the study, which is published with it. The authors of the commentary also emphasize problems with the study methodology.

This case study is one of a number published by the Congress of the USA, Office of Technology Assessment.

17 Henderson, J. B. (1982). An economic appraisal of the benefits of screening for open spina bifida. *Social Science and Medicine* **16**, 545–60.

1. Study design

1.1. *Study question*

What are the net benefits of screening for open spina bifida?(d)

1.3. *Alternatives appraised*

Screening *versus* (implicitly) 'doing nothing'.

1.3. *Comments*

The study concentrated on the benefits of screening. The results were then compared with the costs of a mass screening programme, calculated in a separate study. (See 5 below.)

2. Assessment of costs and benefits

2.1. *Enumeration*

The paper considered tangible benefits to society as a whole from changes in consumption and output (whether marketed or non-marketed) resulting from the prevention of spina bifida births and their replacement with normal births. Intangible benefits which parents derive from children were also estimated. The costs and benefits to the aborted and replacement fetuses themselves were not considered, partly for practical reasons but mainly because the argument becomes one of ethics rather than economics and should be considered separately from, and in addition to, the results of the economic appraisal. The prevention of subsequent spina bifida births to parents with a previous affected pregnancy was not considered as a benefit of the screening programme, as these cases would normally be subject to a prenatal testing.

2.2. *Measurement*

The net benefits were calculated by comparing the net costs to society of a cohort of handicapped individuals, born in the absence of screening, with that of a replacement cohort, which may be smaller or larger in number. The study attempted to identify marginal costs, but where these were difficult to identify average costs were used.

2.3. *Explicit valuation*

Market prices and gross employment costs were used to value tangible benefits, although the author pointed out the sources of possible deviation between these and the appropriate economic values. Public sector goods and services were priced according to official cost data. The costs to parents associated with a normal child were taken as a *minimum* estimate of the psychic benefit to the parents of a normal birth.

3. Allowance for differential timing and uncertainty

Benefits were discounted at 7 per cent (the UK Treasury Test Discount Rate in existence at the time) and also at 4 and 10 per cent to show the sensitivity of the results. Five rates of replacement for aborted handicapped fetuses were used in the paper; 0, 50, 100, 150, and 200 per cent. The delay in replacement was estimated at one year, with additional births, under the 150 and 200 per cent replacement assumptions, delayed by a further two years. The psychic benefit from handicapped children was first assumed to be the same as for a normal child and was then reduced to one-half and to zero, in order to test the sensitivity of the results to this assumption.

4. Results and conclusions

Results were presented for a cohort of 100 spina bifida births, under the different assumptions of the sensitivity analysis, and compared with the screening costs of preventing 100 spina bifida births. The benefit from screening was greatest if no replacement was assumed and declined as the rate of replacement increased. Higher discount rates also reduced the present value of benefits. If parents experienced a loss of psychic benefit through having a spina bifida child, the benefit of screening was increased.

With 100 per cent replacement and assuming that net psychic benefits from a handicapped child were zero, the benefit estimate for a cohort of 100 births prevented was between £1.9 million (4 per cent discount) and £1.1 million (10 per cent discount). The estimated cost of averting these births was £0.3 million. The only set of assumptions for which the benefits were less than the cost of screening was that of 200 per cent replacement with no gain in psychic benefit, discounted at 10 per cent. Taking the most reasonable estimates, the author concluded that 'the results suggest that the tangible benefits of a screening programme probably exceed the tangible costs by about a million pounds per hundred births averted per year'. Allowing for the psychic benefit which may also be derived, this might increase by a further quarter of a million pounds.

5. General comments

The paper includes a survey of previous studies and provides detailed discussion of the methods used in estimating the benefits.

The costs of the mass screening programme are taken from Department of Health and Social Security (1979). *Working group on screening for neural tube defects* (Report). London.

8 Hull, R. D., Hirsh, J., Sackett, D. L., and Stoddart, G. L. (1982). Cost-effectiveness of primary and secondary prevention of fatal pulmonary embolism in high-risk surgical patients. *Canadian Medical Association Journal* **127**, 990–5.

1. Study design

1.1. *Study question*

Is a programme of primary or secondary prevention of fatal pulmonary embolism in high-risk surgical patients more cost-effective than the current (no programme) approach?(c)

1.2. *Alternatives appraised*

Current approach (of waiting for clinical symptoms to appear) *versus* primary prevention (subcutaneous low dose heparin, intravenous dextran or intermittent pneumatic compression of legs) *versus* secondary prevention (leg scanning with ^{125}I-labelled fibrinogen).

1.3. *Comments*

2. Assessment of costs and benefits

2.1. *Enumeration*

The costs considered were the hospital resources used in prevention, diagnosis and treatment of a cohort of 1000 patients assigned to each of the five management strategies. Effects were assessed in terms of the deaths occurring in each cohort. The complications arising from each alternative were mentioned but not quantified.

2.2. *Measurement*

Effectiveness data for heparin, dextran and the current approach were drawn from randomized controlled trials, some from the same geographical location. No evidence on the effectiveness of leg scanning or leg compression was given, although it was argued that these approaches can prevent the onset of deep-vein thrombosis, which is a precursor to pulmonary embolism.

2.3. *Explicit valuation*

Costs were estimated using physicians' fees, and hospital operating costs (excluding treatment costs) and treatment costs for these procedures. (These data were in part drawn from a previous article. See 5 below.)

Hull *et al.* (1982)

3. Allowance for differential timing and uncertainty

Differential timing was not relevant in the context of this study. A sensitivity analysis was conducted, concentrating on possible variations in the costs of prophylactic procedures and hospitalization.

4. Results and conclusions

With the 'no programme' approach, eight patients per 1000 would die at a total cost of Canadian $80 000. Alternatively, primary prevention would save seven lives at total costs varying from $40 000 (heparin) to $135 000 (dextran). Secondary prevention (leg scanning) would probably also save seven lives per 1000 patients, but would cost $350 000. This was mainly because a large number of 'pre-clinical' cases would be identified and then treated by anti-coagulant therapy. Even assuming higher prophylaxis costs and lower hospitalization costs (assumptions unfavourable to primary prevention), primary prevention with low-dose heparin remained the most cost-effective approach.

5. General comments

Some of the cost data for this study were drawn from an earlier paper by the same authors, where more details of cost calculations are given; see Hull, R. D. *et al.* (1981). Cost-effectiveness of clinical diagnosis, venography and noninvasive testing in patients with symptomatic deep vein thrombosis. *New England Journal of Medicine* **304**, 1561–7. (This paper is also reviewed in this volume.)

Hur, D. and Gravenstein, J. S. (1979). Is ECG monitoring in the operating room cost effective? *Biotelemetry Patient Monitoring* **6**, 200–6.

1. Study design

1.1. *Study question*

Is ECG monitoring of all theatre cases during anaesthesia worthwhile?(d)

1.2. *Alternatives appraised*

Routine ECG monitoring during anaesthesia *versus* (implicitly) no monitoring.

1.3. *Comments*

The study was carried out retrospectively. All 7000 non-cardiac anaesthetics in a large, private teaching hospital during one year were reviewed, and the frequency of arrhythmia in this population determined by examination of a sample of 900 consecutive anaesthesia charts.

2. Assessment of costs and benefits

2.1. *Enumeration*

The frequency of life-threatening arrhythmias was monitored from retrospective data, although arrhythmias were not always recorded. Benefits were defined as the number of cases in which ECG monitoring proved 'helpful' in the judgment of attending physicians. The costs used were the highest estimate for ECG monitoring per case, and the lowest estimate of the 'dollar loss' for cardiac arrest with subsequent brain damage. (The objective was to obtain a conservative estimate of the net benefits of routine ECG monitoring.)

2.2. *Measurement*

It was found that in two of the 7000 cases the ECG led to striking and probably life-saving interventions. An absence of monitoring would have led to delayed treatment and probably permanent brain damage. Average costs were used. The cost of ECG monitoring probably included the capital cost of equipment, and re-usable parts. A figure of US$6 per case, quoted by another study, was taken as a high estimate, but it was unclear what was included in this figure. The 'dollar loss' of permanent brain damage was quoted as US$100 000, and was probably derived from malpractice suits, though again it was unclear from the details given.

2.3. *Explicit valuation*

It is assumed that where appropriate, market values were used, but nowhere was it stated explicitly how the costs quoted were derived.

3. Allowance for differential timing

It was implied that capital costs were analysed, but it was not clear how, and what discount rate was used. Although the costs used erred on the side of caution, no range of costs was given. No sensitivity analysis was carried out on the benefit estimates, an important omission given the limited data.

4. Result and conclusions

The authors applied the decision rule that the cost of the diagnostic test should be less than the probability of a mishap times the cost of a mishap. They calculated that the breakeven cost of a test would be US$28.57, and since the cost they quoted was US$6, then ECG monitoring was cost-effective. The authors therefore recommended routine ECG monitoring for all anaesthetics in the United States.

5. General comments

The authors acknowledge the limitations of a retrospective study. In particular they noted that arrhythmias were only recorded if the anaesthetist 'saw them and if he bothered to record them'. The main clinical question was whether arrhythmias would be detected by a good anaesthetist anyway, particularly if they were so serious as to be clinically significant. Also what would be the costs and outcomes of the actions taken to reduce arrhythmia? The study threw little light on these issues, although the authors cite other work.

The authors also discussed the possibility that ECG could be used only for high risk cases; for example, for older patients or those in poor health, or for long operations. However, they pointed out that some studies showed a large proportion of cardiac arrests occurring in otherwise healthy patients.

In general this study was deficient in a number of respects and we would not recommend it as a model for evaluating the economic efficiency of diagnostic procedures.

Koplan, J. P., Schoenbaum, S. C., Weinstein, M. C., and Fraser, D. W. (1979). Pertussis vaccine: an analysis of benefits, risks and costs. *New England Journal of Medicine* **301**, 906–11.

1. Study design

1.1. *Study question*

What are the benefits, risks, and costs associated with the use of pertussis vaccine?(d)

1.2. *Alternatives appraised*

The inclusion of pertussis vaccine in a community-wide vaccination programme *versus* no pertussis vaccine.

1.3. *Comments*

Decision analysis was used to estimate the number of children in a cohort of one million from birth to six years of age experiencing all the possible outcomes.

2. Assessment of costs and benefits

2.1. *Enumeration*

Outcomes included numbers of cases, hospitalization, pertussis-related complications, and deaths. The direct costs of all the outcomes were calculated, including the costs for major and minor vaccine complications. These included the costs of hospitalization and treatment, the costs of residual treatment over time, and the costs of the vaccine.

2.2. *Measurement*

Costs were calculated using cohort analysis. Alternative vaccine strategies were compared for their effects on morbidity, mortality and medical care costs. Probabilities were either based on published data, or on assumptions made by the authors and subjected to sensitivity analysis.

2.3. *Explicit valuation*

Explicit costs were given for all treatment regimens, and were taken from existing publications. It is assumed that market values were used.

3. Allowance for differential timing and uncertainty

Treatment costs incurred over the life of vaccine-damaged children were discounted at 5 per cent. The authors conducted sensitivity analysis by using the lowest and highest rates for all outcome probabilities in the study and for all cost estimates.

4. Results and conclusions

The risks of death and morbidity associated with disease and vaccination were calculated for the cohort with and without the vaccination programme.

This comparison showed that the programme would reduce the costs associated with pertussis by 61 per cent—from US$1 866 153 to US$720 862. The costs of the programme were calculated to comprise US$2194 associated with the disease and US$648 668 with the vaccine (or US$1 226 060 assuming a high cost of vaccine reactions). Using a wide range of probabilities and costs from published material, the benefit–cost ratio only fell below unity in an extreme case.

5. General comments

Only the direct costs were explicitly calculated. If the indirect costs associated with lost future earnings of parents caring for sick children were included, it would not alter the overall results and would tend to favour vaccine programmes.

Most of the costs of the vaccination programme resulted from adverse vaccine reactions, and therefore the authors conclude that the use of less toxic vaccine would make vaccination programmes more attractive.

Kristein M. M. (1980). The economics of screening for colo-rectal cancer. *Social Science and Medicine* **14C**, 275–84.

1. Study design

1.1. *Study question*

What are the costs and benefits of pursuing a haemoccult screening programme for detection of colo-rectal cancer in a specific asymptomatic population?(d)

1.2. *Alternatives appraised*

Haemoccult screening programme *versus* (implicitly) no screening programme.

1.3. *Comments*

The study uses a decision analysis approach. Results are calculated on the basis of screening a population of 100 000 asymptomatic people over the age of 55 years.

2. Assessment of costs and benefits

2.1. *Enumeration*

The total costs of screening the population were calculated, by summing the results at all the decision nodes. Costs only included those of the haemoccult screening procedure and of the confirmatory set of slides. Total benefits included reduced medical care expenditure due to an improvement in the stage of cancer presented, the monetary value of the saved lifetime incomes due to the decrease in case fatality rate, and the gain in working income due to the early identification of the disease. Net benefit was defined as total benefit minus costs.

2.2. *Measurement*

All costs were valued at 1978 prices. The costs of the screening procedures and tests, at US$5 and US$350 respectively, were quoted as being the national average. It was not stated how these costs were calculated or what they included. Each year of life was valued at US$10 000, a figure from the 'low-end' of the human-capital approach. A cost of US$40 was given as the value of each working day gained, but the author does not say how the figure was derived.

Epidemiological data used in the decision analysis and in the valuation of benefits were derived from published studies.

Kristein (1980)

2.3. *Explicit valuation*

It is assumed market values were used, but no details were given.

3. Allowance for differential timing and uncertainty

Results were calculated using a range for the sensitivity and specificity of the screening and acceptance rate, and for different lengths of lead time. Benefits were discounted at 10 per cent per annum. Calculations were not adjusted for radiation risks.

4. Results and conclusions

Given current epidemiological and clinical data, the incidence of colorectal cancer in a population of 100 000 was calculated as 300. The total screening costs were calculated as lying between US$198 000 and US$527 000, the total benefits between US$2 and US$6.6 million, and the net benefits between US$1.8 and US$6.1 million. The benefit cost ratio at 15 years lead time lay between 2.3 and 2.9, and the cost per true positive between US$5500 and US$7000.

However, the direct savings (i.e. reduction in medical care expenditure less costs of screening) were very small or negative, and the author concluded that properly conducted haemoccult screening may be a cost-effective method of early detection of cancer, but that it was essential to increase the long-term survival from its current 40 per cent.

5. General comments

The methodology was very clearly laid out, and the epidemiological and clinical issues carefully considered. However, the costing aspect was very poor. It was unclear how the costs of the screening programme or the values placed on lives were derived. The latter were particularly suspect.

Finally, no other method of screening was considered as an alternative.

·2 Mulley, A. G., Silversteen, M. D., and Dienstag, J. L. (1982). Indications for use of Hepatitis B vaccine, based on cost-effectiveness analysis. *New England Journal of Medicine* **307**, 644–52.

1. Study design

1.1. *Study question*

What would be the most cost-effective strategy for the use of hepatitis B (HB) vaccine?(c)

1.2. *Alternatives appraised*

Vaccinating everyone *versus* screening everyone and vaccinating those without evidence of immunity *versus* neither vaccinating nor screening but passively immunizing those with known exposure.

1.3. *Comments*

A decision analysis approach was used to estimate the likely costs and benefits of alternative vaccination strategies in different populations at risk and: 'to provide a framework for clinicians and policy makers to use in addressing three questions. Who is likely to benefit most from vaccination against hepatitis B virus? Among those who benefit, for whom is vaccination likely to be cost saving? What is the role of screening for prior infection and presumed natural immunity in formulating clinical indications and vaccination policies?'

2. Assessment of costs and benefits

2.1. *Enumeration*

The costs considered were the direct costs of medical resources required to screen and/or vaccinate populations against HB. From these costs were subtracted the medical costs averted by preventing HB infections.

Other potential benefits from vaccination, such as reductions in mortality and morbidity, and averted production losses were noted, but the main measurements concerned the direct costs and savings. The object was to calculate the net cost per case prevented.

2.2. *Measurement*

The efficacy of the vaccine was estimated from a previously conducted controlled clinical trial. Serious adverse reactions were estimated at a frequency of one in 100 000; it was assumed that 10 per cent of these

Mulley *et al.* (1982)

would be fatal. (The costs of treating adverse reactions were considered negligible.) A number of assumptions were made about the natural history of HB infection in order to quantify outcomes. (These are given in an appendix to the paper.) Length of hospital stay for acute hepatitis B was obtained from the literature.

2.3. *Explicit valuation*

The costs of the vaccine (and handling fees) and the costs of treatment were estimated using current medical fees, laboratory charges, and daily hospital charges (1980 prices).

3. Allowance for differential timing and uncertainty

Costs associated with the chronic sequelae of HB infection (up to 20 years after the date of infection) were discounted at 6 per cent.

Sensitivity analysis was performed. The parameters varied were vaccine cost, screening cost, hepatitis associated medical costs, discount rate and vaccine efficacy. Also, most study results (e.g. net costs) were estimated for different annual attack rates.

4. Results and conclusions

It was found that screening followed by vaccination of homosexual men, and vaccination without prior screening of surgical residents, would result in savings in medical costs. Neither screening nor vaccination would be the lowest cost strategy for the general population. Vaccination of susceptible persons would save medical costs for populations with annual attack rates above 5 per cent. (If indirect costs—production savings—were considered this figure may be as low as 1 or 2 per cent.)

The authors pointed out that if averted production losses were considered, even broader indications for the use of HBV would be justified. Furthermore, the central question was: 'How much are we willing to pay to avert the morbidity and mortality as well as the loss of productivity?'

5. General comments

In the discussion section, the different notions of 'benefit' from vaccination were introduced, broadening out from cost savings to willingness-to-pay. The contribution of analysis to such decisions was discussed.

23

Nelson, W. B., Swint, J. M., and Caskey, C. T. (1978). An economic evaluation of a genetic screening programme for Tay–Sachs disease. *American Journal of Human Genetics* **30**, 160–6.

1. Study design

1.1. *Study question*

What are the costs and benefits associated with implementation of a genetic screening programme for Tay-Sachs disease?(d)

1.2. *Alternatives appraised*

Screening programme *versus* treating babies suffering from Tay–Sachs (T–S) disease.

1.3. *Comments*

Although the evaluation was directly applicable to T–S screening programmes, the authors argue that the methodological framework could be used to evaluate other genetic screening programmes.

2. Assessment of costs and benefits

2.1. *Enumeration*

The benefits considered were the averted costs associated with hospitalization of T–S babies. The screening costs included the direct costs of executing the programme plus indirect costs incurred by those tested in the programme. Direct costs included costs of labour, use of personal facilities and materials, and education material. Indirect costs included the value of travel and waiting time. It was assumed that if a T–S baby was identified, pre-birth by amniocentesis, an abortion would be requested. The costs of amniocentesis and abortion were assumed to be the same as the costs of a normal delivery and therefore were omitted from the analysis.

2.2. *Measurement*

Benefits were measured in three stages. First, the average costs associated with the hospitalization of a T–S baby were calculated. Secondly, the impact of the screening programme on the number of T–S babies born was estimated. Thirdly, the expected medical care costs averted due to the discovery of an unmarried carrier, a marriage involving one carrier, and for couples identified as being at risk were calculated. Total costs of 77 carriers were calculated.

Nelson *et al.* (1978)

2.3. *Explicit valuation*

Costs of hospitalization were derived by reference to existing literature, corroborated by data from local hospitals providing such care. Although it was not made explicit, it must be assumed that market values were used to calculate programme costs.

3. Allowance for differential timing and uncertainty

A discount rate of 7 per cent was applied to costs incurred over the T–S baby's lifetime. A range of hospital costs was used, associated with a varying degree of severity of disease. A very conservative estimate was included. Different probabilities were applied to the various categories of risk. However, the authors did not consider the effects of varying the probability that an abortion would be requested if amniocentesis revealed a T–S baby.

4. Results and conclusions

The authors reported their results in two ways. First, they considered the decision rule that the net present value of the benefits of the screening programme must exceed the costs. The conservative, low and high estimates gave the net present value of benefits as US$48 787, US$75 102 and US$150 204 respectively. The net present value of the costs of screening was US$35 000. Therefore, even with a conservative estimate of the benefits, the programme benefits exceeded the costs. Secondly, the authors then report results in terms of benefit: cost ratios for the averted treatment costs *versus* the screening costs. The ratios of 2:1 (conservative), 3.2:1 (low), and 6.4:1 (high) all satisfied the criterion of being greater than unity.

5. General comments

The results only apply to a high-risk population, such as that used in the study. The appraisal did not consider the intangible benefits of the screening programme, only the medical costs averted. However, as the programme was shown to be justified on the basis of tangible benefits, the existence of other benefits is likely to reinforce the result, although the psychic costs associated with abortion should also be considered.

24

Patrick, K. M. and Woolley, F. R. (1981). A cost–benefit analysis of immunization for pneumococcal pneumonia. *Journal of the American Medical Association* **245**(5), 473–7.

1. Study design

1.1. *Study question*

Is immunization against pneumococcal pneumonia justified on the basis of a favourable benefit–cost ratio?(d)

1.2. *Alternatives appraised*

No vaccination programme *versus* a vaccination programme for all adults aged 18 years and over *versus* a programme to identify and vaccinate all high-risk persons.

1.3. *Comments*

The impact of vaccination on the population of a health maintenance organization (HMO) was examined. Decision analysis was used, converted to mathematical expressions for computer analysis.

2. Assessment of costs and benefits

2.1. *Enumeration*

Direct costs were defined as those borne by the providers of the programme, and included costs of vaccine, developing and sustaining the vaccination programme, of treating side effects, and of identifying high-risk populations. Direct benefits were defined as the outpatient and inpatient costs currently borne by the HMO in treating a case of pneumococcal pneumonia that could be averted by the implementation of a vaccination programme. Indirect costs were defined as those borne by the patient entering into the vaccination programme. Indirect benefits were defined as those benefits accruing to patients that would have been forgone if the vaccination programme were not implemented.

2.2. *Measurement*

Clinical and financial data were derived from the HMO's experience and were supplemented by estimates from relevant published studies. Following the decision analysis, average costs per patient were calculated under the three options, based on the outcome probabilities. Projected income loss or gain was used in the calculations as a measure of indirect costs and benefits.

Patrick and Woolley (1981)

2.3. *Explicit valuation*

No information was given in the paper concerning valuation. It must be assumed that market prices were used. The cost of patient time was very low, and it is unclear how this was estimated.

3. Allowance for differential timing and uncertainty

Probabilities of contracting the illness (i.e. as a member of a high- or low-risk population) were applied in the decision analysis. Variables were also applied which took account of the number of years of operation of the programme, the projected growth rate of the Family Health Programme, and the discount rate. Each variable and cost was tested within a range, and the resulting range of benefit–cost ratios under each option was calculated.

4. Results and conclusions

Using direct costs and benefits only, neither of the vaccination programmes yielded a benefit–cost ratio of greater than unity. Sensitivity analysis using wide variations in the values of any particular variable, or set of related variables, had little effect on the ratios.

Using both direct and indirect costs and benefits, the benefit–cost ratio associated with the vaccination programme aimed at all adults was less than unity, and therefore could not be justified. The vaccination programme aimed at the high-risk population did yield a cost–benefit of greater than unity. Sensitivity analysis, as before, had little effect on the ratios, except when varying the incidence of the disease; if disease rates were approximately half those used in the analysis, then costs would be greater than benefits. The implication was that very high-risk groups must be identified. All results relied on high effectiveness rates in high-risk persons.

5. General comments

The authors felt that the cost per case of pneumococcal pneumonia was relatively low, but it is unclear whether this was due to the underlying health of the population (the study was carried out in Salt Lake City), or the method of identification. More costly cases would serve to strengthen the case for vaccination.

The authors also felt it worth noting that the costs were only greater than the benefits when *all* the costs and benefits were taken into account. Therefore, what are the policy implications of this? The doctors operating the HMO would have no incentive to introduce the vaccination programme as it would cost them money. (See also Willems *et al.* 1980, summarized in this section.)

25 Ponnighaus, J. M. (1980). The cost/benefit of measles immunization: a study from Southern Zambia. *Journal of Tropical Medicine and Hygiene* **83**, 141–9.

1. Study design

1.1. *Study question*

Do the benefits of a continuous immunization programme for measles in
a developing country exceed the costs?(d)

1.2. *Alternatives appraised*

Programme of measles immunization *versus* (implicitly) 'doing nothing'.
(See 1.3 below.)

1.3. *Comments*

The paper also considers the relative costs and benefits of programmes
mounted in urban and rural areas.

2. Assessment of costs and benefits

2.1. *Enumeration*

The costs considered included those of the vaccine and its distribution.
The benefits enumerated were the savings in the use of health services,
gains in economic output and satisfaction resulting from better health.

2.2. *Measurement*

The costs of the vaccine and its distribution were estimated based on a
number of assumptions concerning the practicalities of mounting, and
maintaining a 'cold chain', i.e. the system required to ensure that the
vaccine is kept refridgerated until the point of administration. A particular implication was that, since paraffin refridgerators are not reliable
enough, vaccination would have to be carried out by a mobile team in
many areas.

The calculation of benefits centred on savings to the government (e.g.
in reduced utilization of treatment facilities) and did not include patients'
cost savings. A vaccine effectiveness of 0.9 was assumed—that is, at least
10 per cent of the children vaccinated would still get measles.

2.3. *Explicit valuation*

Cost estimates were based on market prices, including current salaries of government employees. Shadow prices were not derived. Market prices were also used to estimate the savings in treatment costs and averted lost production. An average lifetime production loss (of 150 kwacha) was assumed, although the author pointed out that 'ideally this value should be chosen by the society concerned by deciding on the value for the social rate of discount'.

3. Allowance for differential timing and uncertainty

The impact of different discount rates (4 to 15 per cent) on future income streams was explored, although an average value for the present value of production losses was assumed. (See 2.3 above.)

Estimates of costs and/or benefits were made under different assumptions concerning the coverage of the vaccination programme (75 and 100 per cent) and the proportion of measles victims seeking treatment (20 to 80 per cent).

4. Results and conclusions

It was concluded that 'measles immunization programmes should for the time being only be carried out in areas and at centres which have a 24-hour electricity supply', and that 'in rural areas various studies and pilot projects are required to help make rational decisions possible as to what is the cheapest way to reduce infant and toddler mortality rates'.

It was further argued that the extensive use of measles vaccines in towns should give a financial net gain to the government, which could then be used to finance such projects in rural areas. This would counterbalance the bias of health service towards urban areas at the present time.

5. General comments

This study is particularly strong in its discussion of the ways in which practical problems in vaccine administration affect programme costs. It is also one of the few studies to address the urban *versus* rural health service delivery problem in developing countries. However, some of the issues raised, in particular the equity-efficiency trade-off, are only partially dealt with.

A number of the methodological issues arising in the estimation of the costs and benefits of immunization programmes are discussed in Creese, A. L. and Henderson, R. H. (1980). Cost–benefit analysis and immuniza-

97

Ponnighaus (1980)

tioin programmes in developing countries. *Bulletin WHO* **58**(3), 491–7. A problem pertaining to this study, in common with many undertaken in developing countries, is that of assessing the contribution to life expectancy from the programme concerned, given the high risk of death from other causes.

26 Rosenstein, M. S., Farewell, V. T., Price, T. H., Larson, E. B., and Dale, D. C. (1980). The cost-effectiveness of therapeutic and prophylactic leukocyte transfusion. *New England Journal of Medicine* **302**, 1058–62.

1. Study design

1.1. *Study question*

What is the effectiveness and cost of leucocyte transfusion in the treatment of patients with acute leukemia?(c)

1.2. *Alternatives appraised*

Therapeutic leucocyte transfusion *versus* prophylactic leucocyte transfusion.

1.3. *Comments*

The study considered the incremental costs and effectiveness for prophylactic transfusions *vis-à-vis* therapeutic, as well as the simple cost-effectiveness ratios.

2. Assessment of costs and benefits

2.1. *Enumeration*

Only the direct costs of treating patients were included. Direct costs were for leukapheresis and hospitalization. Effectiveness was expressed in terms of life years saved.

2.2. *Measurement*

The effectiveness of leucocyte transfusion was calculated by means of an odds–ratio technique yielding probabilities of mortality in patients with transfusion and similar patients without.

Using an incidence rate for leukemia of 11 000 new cases per year, each receiving an average 1.5 episodes of treatment, of which 65 per cent require transfusion, potential national annual costs for transfusion were calculated. The study was linked to a controlled clinical trial.

2.3. *Explicit valuation*

Market values and normal fees and charges were used for the cost of leukapheresis and hospitalization.

Rosenstein *et al.* (1980)

3. Allowance for differential timing and uncertainty

Using sensitivity analysis, a range of values for the cost-effectiveness ratio (in incremental cost per life year gained) was calculated by varying (one at a time) the probability of infection or death in the group given transfusion, and the cost per granulocyte transfusion.

4. Results and conclusions

Therapeutic transfusion was anticipated to reduce the probability of mortality in selected patients from 36 to 15 per cent and overall mortality from 23 to 10 per cent. Prophylactic transfusion was anticipated to reduce the probability of documented infection per illness episode from 65 to 28 per cent. Expressed as life years gained per case of treatment the range was 0.07 to 0.11.

The cost per leukapheresis ranged from US$80 to US$300. The average cost of therapeutic transfusion was calculated to be US$1648 and of prophylactic transfusion to be US$3502, to be added on to the average cost of US$9864 for hospitalization. On the basis of the known incidence of acute leukemia, the national per annum costs for therapeutic and prophylactic leucocyte transfusion were US$17.7 million and US$57.8 million respectively. The cost-effectiveness ratio for therapeutic leucocyte transfusion was an average of US$14 982 per life year gained. Prophylactic transfusion cost an average of US$50 029 or US$35 020 per life year gained (depending on the effectiveness of continuing the transfusions). The cost per additional life year gained via prophylactic rather than therapeutic transfusion (assuming maximum efficacy) would be approximately US$85 281.

5. General comments

This technique is useful for examining new and expensive treatment methods. Effectiveness was measured by reference to decreased mortality only, and not decreased morbidity. The authors offer suggestions on how leucocyte transfusion could be made more cost-effective.

27 Shapiro, M., Schoenbaum, S. C., Tager, I. B., Muñoz, A., and Polk, F. (1983). Benefit–cost analysis of antimicrobial prophylaxis in abdominal and vaginal hysterectomy. *Journal of the American Medical Association* **249**(10), 1290–4.

1. Study design

1.1. *Study question*

Is it worthwhile prescribing antimicrobial prophylaxis routinely for prevention of wound infection after hysterectomy?(c) (See 1.3 below.)

1.2. *Alternatives appraised*

Antimicrobial prophylaxis (cefazolin sodium) *versus* no prophylaxis, following vaginal or abdominal hysterectomy.

1.3. *Comments*

A decision tree was used to set out the sequence of clinical alternatives and outcomes for both groups.

The study is classified as (c) since it essentially poses the question 'what is the most efficient way of caring for this group of patients?'

2. Assessment of costs and benefits

2.1. *Enumeration*

The benefits considered were those health services resources that would be averted by preventing post-operative morbidity (both during hospitalization or within six weeks after discharge). The morbidity considered included infection at the operative site, urinary tract infection or febrile morbidity. These averted costs were compared with the costs of prophylaxis. The pain and discomfort in administration of the drug was acknowledged but not quantified.

2.2. *Measurement*

The study used data from a large, randomized, placebo-controlled, clinical trial of prophylaxis among women undergoing elective hysterectomy. (Vaginal hysterectomy, n = 86; abdominal hysterectomy, n = 429.)

2.3. *Explicit valuation*

Estimates of costs attributable to infectious morbidity were derived from 1981 charges at the Brigham and Womens Hospital, Boston. In general, market prices were used to value costs and benefits.

Shapiro *et al.* (1983)

No value was attached to the avoidance of morbidity *per se*, nor to the unpleasantness associated with prophylaxis (intramuscular injection).

3. Allowance for differential timing and uncertainty

Since all costs and effects occurred in a short period of time, discounting was not relevant in the context of this study.

A sensitivity analysis was performed to project the effect of using newer (more expensive) cephalosporins and of administering prophylaxis for durations longer than those currently recommended.

4. Results and conclusions

In patients undergoing vaginal hysterectomy, prophylactic cefazolin reduced in-hospital infectious morbidity from 52 per cent to 23 per cent; for abdominal hysterectomy it was reduced from 43 per cent to 25 per cent.

It was estimated that patients with post-operative morbidity would both have longer inpatient stays and a higher average daily cost of hospitalization (owing to laboratory tests and administration of antibiotics).

Taking into account the costs of prophylaxis, the average net benefit was US$492 per patient for vaginal hysterectomy and US$102 per patient for abdominal hysterectomy.

These benefits would be eroded by use of newer, more expensive cephalosporins unless these were considerably more effective than cefazolin. The benefits would also be diminished by inappropriate prolongation of the duration prophylaxis. (It was pointed out that routine practice may be four to five days and not the eight to 24 hours used in clinical trials.)

5. General comments

This paper was subject of an editorial in the same edition of the *Journal of the American Medical Association*, p. 1328. Although generally supportive, the editorial points to the dangers of extrapolating from one setting to another. However, it may be that one of the article's strengths is that it might lead physicians in other locations, currently administering prophylaxis for longer periods, to re-examine their practice.

Stason, W. B. and Weinstein, M. C. (1977). Allocation of resources to manage hypertension. *New England Journal of Medicine* **296**, 732–9.

1. Study design

1.1. *Study question*

How can resources be used most efficiently within programmes to treat hypertension?(c)
To what extent does treatment of hypertension pay for itself?(d)

1.2. *Alternatives appraised*

Screening and treatment for hypertension *versus* no organized screening programme. (See 1.3 below.)

1.3. *Comments*

Although the alternatives above represent the basic alternatives examined, the paper considered a range of options and explored the costs and benefits by age at initiation of therapy, sex, and pre-treatment diastolic blood pressure. (See 5 below.)

2. Assessment of costs and benefits

2.1. *Enumeration*

The costs considered were the health sector costs of the screening and treatment programme, the cost savings from the reduction in cardiovascular morbid events, the costs of treating side effects and extra costs of treating non-cardiac illness in added years of life. Patients' costs and benefits, and changes in productive output were not considered. The effects considered related to the changes in life expectancy arising from the programme. Additional life years were adjusted for quality so as to encompass avoided morbidity (as opposed to mortality) and the side-effects of treatment.

2.2. *Measurement*

Few details were given of the measurement of costs or effects in physical units, although details were given of the number of physician visits, investigations and length of hypertension treatment assumed in the calculations. Mortality and morbidity benefits from hypertension control were obtained from the Framingham Study by assuming that reduction of blood pressure (BP) confers some fraction of the benefit implied from the schedules of risk associated with untreated BP. The analysis was performed under three assumptions of varying degrees of conservatism; that risk is reduced to that associated with post-treatment BP, that the risk reduction is only half this amount, or that the risk reduction depends on age at initiation of therapy and the length of treatment.

2.3. Explicit valuation

Few details were given of cost estimation, but in the main health service expenditures were used. A quality adjustment was made to the years of life gained by hypertension therapy in order to reflect treatment side-effects. Two assumptions were made (*i*) that a year of life with side-effects is equivalent to 0.99 of a healthy year, or (*ii*) that it is equivalent to 0.98 of a healthy year.

3. Allowance for differential timing and uncertainty

Costs and benefits occurring in the future were discounted at rates of 0, 5 and 10 per cent. Sensitivity analysis was a central feature in the study. Key parameters varied were the discount rate, the level of adherence to therapy and the value of a year of life with side-effects.

In addition results were presented, in cost per quality-adjusted life year, for males and females, age at initiation of therapy and pre-treatment diastolic BP.

4. Results and conclusions

A major finding was that, within the field of hypertension control, an intervention to improve patient adherence may be a better use of limited resources than maximum efforts to detect hypertension. Public programmes to screen for hypertension were indicated, on cost-effectiveness grounds, only if adequate resources were available to ensure that detection is translated into effective long-term BP control.

Assuming patient adherence to therapy, the cost per quality-adjusted life year gained was US$4850 for patients with BP over 105 mmHg and US$9880 for those with BP between 95 and 104 mmHg. When the problem of patient adherence was introduced, these figures rose to US$10 500 and US$20 400. The authors concluded: 'Are these still reasonable prices to pay? In the absence of comparable analysis for other uses of health-care resources, the answer to this question depends solely on the subjective valuation that one wants to place on a year of life.'

Stason and Weinstein (1977)

5. General comments

The methodology and results reported in this paper are set out in more detail in a longer study: Weinstein, M. C. and Stason, W. B. (1976). *Hypertension: a policy perspective.* Harvard University Press, Cambridge, Mass.

Although slightly older than most of the other studies reported in this volume, it is included because of its prominence and the fact that the cost-effectiveness model used has been extensively referred to by other authors. (The model is set out in full in an accompanying methodological paper: Weinstein, M. C. and Stason, W. B. (1977). Foundations of cost effectiveness analysis for health and medical practices. *New England Journal of Medicine* **296**, 716–21.) The distinctive features of this particular approach to cost-effectiveness analysis are the *exclusion* of costs other than the resource costs to the health sector (e.g. production losses and gains) and the *inclusion* of the treatment costs of non-cardiovascular diseases in added years of life. See also the editorial comment by Fein, R. (1977). *New England Journal of Medicine* **296**, 751–3.

29 Thompson, M. S. and Cohen, A. B. (1981). Decision analysis: electronic fetal monitoring. In *Methods for evaluating health services* (P. M. Wortman, ed.). Sage Publications, New York.

1. Study design

1.1. *Study question*

What are the costs and benefits of electronic fetal monitoring (EFM)?(d)

1.2. *Alternatives appraised*

Electronic fetal monitoring *versus* (implicitly) doing nothing.

1.3. *Comments*

The comparison with the 'doing nothing' alternative was not explicitly recognized. It was not clear, therefore, whether the costs and benefits identified accurately reflected this comparison.

2. Assessment of costs and benefits

2.1. *Enumeration*

Benefits were defined as lives saved and morbidity prevented. Costs were defined as the costs of additional morbidity and mortality expected, and the costs of monitoring.

2.2. *Measurement*

This study was not linked to a clinical trial. Estimates of clinical efficacy and values of morbidity and mortality were taken from other studies. Using the former, probabilities of clinical benefits and risks were calculated. Using the latter, in combination with cost estimates, net costs and benefits were calculated with upper and lower limits.

The costs of the additional morbidity were measured by treatment costs, multiplied by the outcome probabilities. The total costs and benefits associated with using EFM on a cohort of 1000 average births were calculated.

2.3. *Explicit valuation*

The value of a life saved was taken to be the discounted earnings forgone, and the value of a retarded life as discounted earnings forgone plus treatment costs.

The values used were taken from Banta and Thacker, 1979 (summarized in this volume) and other sources in the literature. No indication was given of how these costs were derived or what they represent.

3. Allowance for differential timing and uncertainty

Different estimates of uncertainty and risk were taken into account by applying various rates of perinatal mortality, brain damage, maternal and neonatal infections. A range of upper and lower limits was presented. Interviews were carried out with obstetricians to obtain their estimates of risk.

Expected lifetime earnings were discounted at 10 per cent.

4. Results and conclusions

Using Banta and Thacker's costs, total benefits were estimated to be US$156 400, and total costs US$163 300, giving a net cost per 1000 births of US$11 900. The authors point out that there was considerable uncertainty over the expected reduction in perinatal mortality, and that the magnitude of this uncertainty was too great to allow these results to be used as an aid to decision making. Whilst the methodology was considered acceptable, the authors highlight the need for further research.

5. General comments

This study comprises a thorough survey of the clinical data, but is inadequate as an economic evaluation. In particular, there is little discussion of costing methods. Only average costs were used and these were not broken down into component parts.

30 Veale, A. M. O. (1980). Screening for phenylketonuria. In *Neonatal screening for inborn errors of metabolism* (H. Hickel, R. Guthrie and G. Hammerso, eds). Springer-Verlag, Heidelberg.

1. Study design

1.1. *Study question*

Is it worthwhile screening for and treating phenylketonuria (PKU) in newborns? (d)

1.2. *Alternatives appraised*

Screening and treatment *versus* no screening. (See 1.3 below.)

1.3. *Comments*

Although the paper was primarily concerned with screening and treating PKU, it discussed the possibility that other tests may sometimes be added to a programme and that these, if they involve significant extra costs, should also be appraised.

2. Assessment of costs and benefits

2.1. *Enumeration*

The costs considered were the health sector costs in mounting and running the screening programme (initial equipment, collection of specimens, running expenses and investigations) and in providing 10 years' treatment at home. (The costs in terms of disruption in the home were also mentioned.) The benefits considered were the averted costs of custodial care for handicapped children and the recovery of a potentially productive member of society.

2.2. *Measurement*

Data were based on the operation (over 10 years) of one screening programme in New Zealand. On average 2.63 cases of PKU per year were assumed. Averted institutional costs were estimated using a life table for untreated PKU cases. In the absence of precise information a few assumptions were made.

2.3. *Explicit valuation*

Market values were used to estimate programme costs and the benefits in terms of health service resources saved. No attempt was made to value the production gains or costs and benefits to the family. The issue of whether averted institutionalization for PKU patients actually leads to savings in expenditure, or enables other needs to be met, was discussed.

3. Allowance for differential timing and uncertainty

Costs and benefits occurring in the future were discounted at 10 per cent.

Averted institutionalization costs were estimated under five different assumptions about the mortality experience of PKU sufferers. This was shown to affect the ratio of benefits to costs, the variation being from 1.4 to 0.8.

4. Results and conclusions

Screening for PKU was shown to be worthwhile in economic terms. Imputing a value for improved health state *per se* would only improve the benefit side of the equation.

On the other hand, the author stressed the importance of not adding other elements to the screening programme without economic evaluation.

5. General comments

31 Willems, J. S., Sanders, C. R., Riddiough, M. A., and Bell, J. C. (1980). Cost-effectiveness of vaccination against pneumococcal pneumonia. *New England Journal of Medicine* **303**(10), 553–9.

1. Study design

1.1. *Study question*
Is vaccination against pneumococcal pneumonia worthwhile?(d)

1.2. *Alternatives appraised*
Vaccination against pneumococcal pneumonia *versus* no vaccination.

1.3. *Comments*
The authors carried out two analyses, one relating to all patients, the second to only those people covered by Medicare. The authors used a hypothetical vaccination programme and applied a simulation model for the period 1975 to 2050, comparing the results from the vaccinated and non-vaccinated populations.

2. Assessment of costs and benefits

2.1. *Enumeration*
The costs considered were those incurred in the medical care sector only; net medical care costs included the added cost of vaccination, the reduced costs of treatment of pneumonia, the added costs of treatment of the vaccine's side effects, the added costs of treatment due to extended years of life.

Benefits were expressed as changes in health status, measured in quality adjusted life years. Net effects (measured in QALYs) included increased life expectancy from prevention of pneumonia, the improved quality of life from prevention of morbidity, the reduced quality of life from morbidity due to the vaccine's side effects, and the reduced quality of life from morbidity not prevented by the vaccine.

The cost-effectiveness ratio was expressed as the net medical cost per year of healthy life gained by the vaccinated person.

2.2. *Measurement*
All costs were given at 1978 price levels. The cost per day of hospitalization for pneumonia was that reported by Blue Cross. The costs of vaccination given by a private physician, the costs of treating a patient with pneumonia in an ambulatory setting (including physicians' fees and

diagnostic tests), and the costs of severe systemic reaction as a side effect of the vaccine (including physician's fees and two days in hospital) were taken from published studies.Epidemiological data, and costs of treating other disease were also taken from published studies.

2.3. *Explicit valuation*

It is assumed market values were used. Production losses and gains were not valued, nor were non-medical care costs.

3. Allowance for differential timing and uncertainty

All results were computed taking account of ranges of epidemiological data and variation in costs. Discount rates of 0–10 per cent were applied. Relative values of different states of health were applied, and these QALY weights were varied to test their importance.

4. Results and conclusions

The cost-effectiveness ratios for the base cases varied from US$77 000 per QALY gained for the two-to-four year-age group to US$1000 for the 65-plus age groups. With conservatively estimated vaccination rates equal to the 1975 rates for 'flu vaccination, there would be a net cost of US$23 million to vaccinate the 65-plus age group, yielding 22 000 QALYs; and for all age groups, a net cost of US$150 million would yield 31 000 QALYs. A sensitivity analysis showed that the results can be dramatically altered, depending on the assumptions made. The authors suggested that it is cost-effective to vaccinate high-risk people, but, apart from older people, it is difficult to define high-risk people, and therefore they suggested the need to develop an accurate cost-effectiveness model for a high-risk population, including, for example, those with chronic diseases.

5. General comments

The authors discussed the relevance of their results to third-party insurers, particularly Medicare. Given that it appeared to be cost-effective to vaccinate the elderly, who receive Medicare, it was suggested that it would be beneficial for Medicare to invest in a vaccination programme. However, Medicare excluded preventive services from its reimbursements. This anomaly was highlighted together with the problems of access and utilization of preventive services by the elderly. (See also Patrick and Wooley, 1981, summarized in this section.)

32 Wilson, C. B. and Remington, J. S. (1980). What can be done to prevent congenital toxoplasmosis? *American Journal of Obstetrics and Gynaecology* **138**(4), 357–63.

1. Study design

1.1. *Study question*

Is a screening programme to prevent congenital toxoplasmosis worthwhile?(d)

1.2. *Alternatives appraised*

Screening (and treatment) *versus* (implicitly) doing nothing.

1.3. *Comments*

Alternative methods of preventing congenital toxoplasmosis were discussed in the text but were not pursued further, e.g. improved hygiene. Therefore the study was not really addressing the question of how best to prevent congenital toxoplasmosis (if at all), but simply considered the merits of a screening programme.

2. Assessment of costs and benefits

2.1. *Enumeration*

The costs of screening tests were considered, but not the induced costs of treatment or abortion. The benefits from reduced morbidity were taken to be the avoided costs of special care for affected children. (See the companion volume, Section 2.2.) Additional acute care and the effect on lifetime earnings were not considered.

2.2. *Measurement*

The details of the screening programme costs were not given and it is not clear what was included in the figures quoted: for example, whether labour costs were included and how these were measured. The cost estimate for the programme was based on the assumption that all women were screened.

There is some uncertainty about the incidence of infection amongst pregnant women and the subsequent incidence of congenital toxoplasmosis. Figures quoted in the paper were taken from the available literature and a small-scale follow-up of 13 children who were asymptomatic at birth. No reference was made to marginal costs although there are choices in the timing and frequency of screening tests which may result in different costs per case detected.

2.3. *Explicit valuation*

Cost estimates for special care were taken from a study of congenital rubella in the United States and costs experienced in California. No details of any costings were given but they can be assumed to be based on market prices. No attempt was made to place a value on improved health *per se*.

3. Allowance for differential timing and uncertainty

Despite the fact that a comparison was needed between the costs of screening *now* and benefits to be derived *in the future*, no discounting was employed.

The authors incorrectly use the term 'discounting' to mean the deflation of costs in one year to the prices current in an earlier year.

Although many of the factors involved were uncertain (e.g. incidence of infection, incidence of congenital toxoplasmosis, effectiveness of treatment, take up rate for screening, etc.) no attempt was made to apply sensitivity analysis.

4. Results and conclusions

The average cost of caring for an individual with congenital toxoplasmosis was US$67 246 per lifetime. Given the incidence assumptions in the paper, the total lifetime cost of infants born in one year was US$221.9 million. The cost of a screening programme was estimated to be US$110.7 million per year and it was assumed that 50 per cent of the cases of congenital toxoplasmosis could be prevented. The authors concluded that a screening programme for congenital toxoplasmosis may prove cost-effective and suggested regional screening programmes be set up to provide additional data.

5. General comments

This study was poorly structured and illustrates the pitfalls of failing to apply the framework of economic appraisal rigorously, particularly in circumstances where knowledge is limited. An example of the problem is that the authors recommend that primary prevention (via health education) should be undertaken without having examined the issue at all and also failing to take account of the impact that effective health education would have on the cost-effectiveness of screening.

SECTION 3: Alternatives in diagnosis

Introduction

This section examines economic evaluation of diagnostic procedures, where diagnosis is defined as the process by which problems or diseases are identified. As stated in the introduction to Section 2 there is an obvious area of potential overlap between diagnosis and certain preventive activities. The distinction which has been drawn between diagnosis and the use of the tests for screening, for example, is that diagnostic procedures are used to validate a clinical hypothesis about the cause of symptoms which are presented by the patient, or as part of the clinical work-up to determine the management of the patient.

Of the 11 studies summarized in this section, the majority (eight) are concerned with the relative cost-effectiveness of alternative diagnostic techniques or strategies. Three of the studies (Dixon *et al.* 1981; Evens and Jost 1977 and Larson *et al.* 1980) compare computed tomography (CT) with other diagnostic tests. A further three studies consider alternative diagnostic strategies for deep-vein thrombosis (Hull *et al.* 1981), tuberculosis meningitis (McNeil *et al.* 1980), and pheochromocytoma, a rare cancer which is present in some hypertensive patients (Weinstein and Fineberg 1978). Finally Doberneck (1980) examines alternatives to breast biopsy for the diagnosis of cancer or other complaints and Holmin *et al.* (1980) consider the routine or selective use of an investigative operation, cholangiography.

Three studies attempt to proceed further and look at whether the diagnostic activity is worthwhile *per se*. The procedures evaluated in this way are routine paediatric pre-operative chest X-rays (Neuhauser 1978), routine skull X-rays folllowing head injury (Royal College of Radiologists 1981) and CT scanning of headache patients to detect subarachnoid haemorrhage (Knaus *et al.* 1981). However, these studies fall short of being complete cost–benefit analyses, generally expressing their results in terms of cost per life or life year gained.

114

Particular methodological problems in this area

The studies of diagnostic procedures raise a number of methodological problems some of which have counterparts in other areas but which present differently in the context of diagnosis. Given the similarities between diagnostic activities and some preventive measures, the methodological comments in the introduction to Section 2 may also be relevant here. Also the general comments about average *versus* marginal costing of reductions in hospital stay made in Chapter 2 are also applicable here.

Outcome measures are particularly important in studies of diagnostic techniques because the process is one step removed from treatment and final outcome, i.e. improved health. Whilst it may be argued that information, as such, has an intrinsic value, inability to act upon the information derived from a diagnostic procedure must raise some doubts about its value and if the procedure is invasive it may be considered unethical. Although there is no reason why cost-effectiveness studies should take a narrow view of effectiveness, the bias towards cost-effectiveness rather than cost–benefit analysis has tended to result in evaluative studies more often considering diagnostic efficacy and costs rather than the effect on outcome or the relationship between the cost of the procedure and the information obtained.

This approach was adopted in most of the cost-effectiveness studies reviewed here (and one of the cost–benefit studies), although some did consider final outcomes (Holmin *et al.* 1980; McNeil *et al.* 1980; Weinstein and Fineberg 1978). Cost-effectiveness studies may rely on more restricted outcome measures provided one of the diagnostic procedures is better than all others on at least one criteria of cost, accuracy or safety and no worse on all other criteria. However, some doubt must be cast upon the use of measures such as 'cost per case detected', even for cost-effectiveness comparisons, where this condition does not hold.

A related concern is the quality of clinical evidence upon which outcome or effectiveness measures are based. Several of the studies reviewed in this section are not linked to controlled clinical trials but rely on comparisons with practices in other hospitals (Doberneck 1980; Evens and Jost 1977) or retrospective data from case notes (Holmin *et al.* 1980). These approaches make the clinical data less reliable and make it difficult to identify correctly the marginal impact of proposed alternatives. Some of the authors acknowledge these difficulties but in some cases the appropriate caveats are missing.

The use of marginal analysis has also given rise to problems in the diagnosis studies. In several cases, the results are reported in terms of the average cost per case detected (Doberneck 1980; Evens and Jost 1977). However, as Hull *et al.* (1981) draw out in their study, the proper way to assess procedures of varying diagnostic accuracy is to consider the incremental cost per additional case detected. (See Chapter 2 for a detailed discussion of this point.)

The final area of concern in this section is the handling of capital costs. Several of the studies are concerned with the introduction of new technology, often involving large capital outlays. As indicated in Chapter 2, Section 2.3, there has been a trend towards annuitizing capital outlays rather than discounting future revenue costs. This procedure is perfectly acceptable provided it is carried out correctly. However, some of the studies use straight line depreciation to calculate an annual equivalent cost (for example, Evens and Jost 1977). This is the same as not discounting at all. Other studies ignore discounting and annuitization completely (Dixon *et al.* 1981; Knaus *et al.* 1981). Finally, there is a general tendency to ignore the impact of throughput on unit capital costs. Weinstein and Pearlman (Congress of the USA 1981) addressed this issue in the context of automated equipment for laboratory use and concluded that: 'paradoxically, the cost-effectiveness of the multichannel analyzer may rest on whether we, as a society, are willing to pay enough for diminishing incremental health benefits to justify sufficient volumes of testing to permit us to afford the reduction in unit testing costs that multichannel analyzers offer'. With some notable exceptions (e.g. Evens and Jost 1977) the authors of studies involving expensive capital have not considered what throughput levels would be justified when calculating the capital cost component of test procedures.

Current state of the art

One of the most disappointing aspects of the current work in this area is the lack of attention paid to the essentially probabilistic nature of diagnostic activity. Whilst some studies do employ decision trees and decision analysis to assist their appraisal (see, for example, McNeil *et al.* 1980) in general the results are not couched in terms of expected outcomes. The economic issues which result, such as whether expected outcomes have the same value as a certain equivalent, are untouched by the studies. The question of how much improvements in diagnostic accuracy are worth, and to whom, is largely unexplored. However, the potential conflict between the doctor's interests (in eliminating the remote possibility of a rare underlying condition) and the patient's interests (having to bear the physical and sometimes financial costs of additional investigations) is central to any analysis of efficient and effective diagnosis.

A second area of concern regarding the current literature in this field is the choice of alternatives against which new techniques are to be evaluated. Some studies recognize their limitations in the range of alternatives considered (e.g. Holmin *et al.* 1980). However, in some cases it is not clear whether the alternatives chosen represent *best* alternative practice. The study by Hull *et al.* (1981) is a good example of an evaluation which considered a range of alternatives, including combinations of diagnostic tests, rather than comparing existing practice with the new diagnostic 'gold standard'.

Finally, the coverage of potential issues in this area is also rather disappointing. There has been a clear tendency to focus appraisal activity on the substitution of new techniques, and particularly high technology; four of the 11 studies are concerned with computed tomography (CT) scanning. The relative cost-effectiveness of activities where new developments have not taken place and considerations of how worthwhile are existing diagnostic activities, have tended to be neglected areas. Where such studies have been carried out (e.g. Neuhauser 1978), the results have shown that some of the existing practices are certainly of dubious value.

Contribution to decision making

Methodological weaknesses, particularly in costing, undermine the direct usefulness of many of the studies to decision making. Attention has usually been drawn to this problem in the general comments section of each summary. However, even where the study itself is well conducted, there is a danger that conditions will not be replicated in practice. Tests which have been appraised as *alternatives* may be used as complements, particularly where clinicians are more familiar with an older technique which is being replaced. Savings on reduced use of other tests may not be realized, therefore. Whilst this is not necessarily a problem with which evaluators should concern themselves, it is an area where decision makers' attention needs to be drawn to the problem of controlling clinicians' ordering of diagnostic tests.

Weinstein and Fineberg (1978) highlight the role that decision makers' values play in the final choice of diagnostic strategy. Their presentation and discussion of the results is extremely useful and contrasts with the tendency of other studies to report *the* cost-effective solution. They make the point that in presenting cost-effectiveness results in terms of additional costs per case detected, the chosen strategy should be the one where this cost equals the value placed on the detection of additional

117

cases and not necessarily the strategy with the lowest cost per case detected.

Finally, the studies of new technology, in particular CT scanning, have neglected the broader policy implications, such as the location of scanners. Vital questions about the appropriate uses of equipment, in relation to case mix, the size of potential patient groups and the effect of changing throughput on unit costs have all been neglected. In all issues relating to expensive capital investments, the problem of the appropriate scale of operation needs to be addressed and the trade-off between economies of scale and travel costs for patients considered. (See the study of regional centres for health surgery, Finkler 1981, in Section 6.) The studies here offer no guidance to decision makers on this point and have generally neglected the issues of costs to patients.

Reference

Congress of the USA (1981). Office of Technology Assessment. *The implications of cost-effectiveness analysis of medical technology. Case study 4: Cost-effectiveness of automated multichannel chemical analyzers.* Washington, D.C.

3 Dixon, A. K., Fry, I. K., Kingham, J. G. C., McLean, A. M., and White, F. E. (1981). Computed tomography in patients with an abdominal mass: effective and efficient? *Lancet* **i**, 1199–1203.

1. Study design

1.1. *Study question*

What are the costs and benefits associated with different methods of diagnosing an abdominal mass?(c)

1.2. *Alternatives appraised*

Conventional imaging *versus* computed tomography (CT).

1.3. *Comments*

The objective was to identify the most successful, least uncomfortable, and most cost-effective diagnostic sequence. The two procedures were compared on the basis of their use as the initial diagnostic test.

2. Assessment of costs and benefits

2.1. *Enumeration*

The costs associated with two patterns of diagnosis were calculated (i.e. the time taken to reach a diagnosis, associated investigations, and stay in hospital). The subsequent pattern of treatment costs was not considered. It was not clear whether or not capital costs were included. Benefits were defined in terms of reduced costs, clinical benefits, the accuracy of the imaging, and the reduction in risks associated with unnecessary investigations.

2.2. *Measurement*

The study was linked to a randomized controlled trial. Sixty patients were allocated between two groups, with CT scanning being the first diagnostic procedure used for one group. The other group was investigated by available alternatives. Outcome was measured by seven factors including the time taken to diagnosis, the need for inpatient investigation, the number of inpatient days needed to reach a diagnosis, the number and risks of imaging investigations, the cost of imaging, the need for laparotomy, and the accuracy of imaging. These factors were followed up for a minimum of nine months. The accuracy of the clinician's examination and diagnosis were also assessed. Costs were

119

measured as average costs rather than marginal. Only the average cost per diagnosis was calculated and not the cost per positive result.

2.3. *Explicit valuation*

The costs of all investigations and imaging were calculated by reference to private radiological establishments and the costs per in-patient day were assessed at private clinic rates.

3. Allowance for differential timing and uncertainty

Differential timing was not considered. Discounting (or annuitization) would be relevant in comparing capital costs for different diagnostic procedures. In particular, the scanning option may require capital equipment purchases sooner rather than later.

No sensitivity analysis was carried out although there were factors for which it would have been relevant. In particular, the level of patient throughput would affect both average and marginal costs, but the authors did not discuss the throughput assumptions underlying the cost figures.

4. Results and conclusions

The time taken to make a diagnosis, the length of stay in hospital, and the number of investigations for the patients given CT were all less than for the conventional imaging group. Although the cost of the CT itself was higher than the cost of conventional imaging, the total cost, including inpatient stay, was lower for the CT group than for the conventional imaging group. The difference in the cost of imaging was small and the authors argued that the cost of CT will fall as the number of additional tests are reduced and if CT examinations are tailored to fit patients. Most of the difference in cost between the two groups was associated with the length of stay. No significant differences in the accuracy of imaging were observed between the two groups.

5. General comments

The results should be considered with some caution. Since length of stay between the two groups is a crucial factor in the difference in costs, it is necessary to examine the alternative uses to which the 'spare' bed days could be put, if any, in order to establish their opportunity cost (see the companion volume, Section 7.1.2). Also, the use of private sector costs implies that the results cannot necessarily be assumed to be applicable in NHS circumstances. It is also unclear whether such costs include a capital element.

34 Doberneck, R. C. (1980). Breast biopsy: a study of cost-effectiveness. *Annals of Surgery* **192**, 152–6.

1. Study design

1.1. *Study question*

What is the cost-effectiveness of breast biopsy for patients with a diagnosis of breast pain, fibrocystic disease, or breast neoplasm?(c)

1.2. *Alternatives appraised*

(*a*) Alternative approaches to breast biopsy, involving general anaesthesia, assisted local anaesthesia and local anaesthesia. (*b*) Breast biopsy *versus* mammographic screening *versus* judicious use of mammography in patients with breast complaints.

1.3. *Comments*

2. Assessment of costs and benefits

2.1. *Enumeration*

The costs considered were those relating to hospital resource use only. This included operating/recovery room charges, electrocardiogram and other laboratory tests, room charges and surgeons' fees. Benefits in terms of long-term outcome were not considered, although the number of cancers detected was recorded.

2.2. *Measurement*

Data on breast biopsies performed, anaesthetic practice followed, cancers detected and costs incurred were obtained from the hospital under study over a six year period. These were then compared with data from other locations. The study was not linked to a controlled clinical trial.

2.3. *Explicit valuation*

Market values were used to estimate costs.

3. Allowance for differential timing and uncertainty

Not considered. Discounting might have been relevant in the comparison of breast biopsy with mammography as resource outlays are likely to occur at different times.

Doberneck (1980)

4. Results and conclusions

It was found that the cost per cancer detected (in the hospital under study) was US$6411. However, biopsies performed with assisted local anaesthetic were much cheaper than those performed with general anaesthetic (US$751 *versus* US$1216). Greater precision in selection of patients for biopsy reduced the cost of diagnosing cancer. (A method for doing this is discussed.) However, it was argued that the present state of the surgical art was more cost-effective than was screening for breast cancer.

5. General comments

The paper discusses the savings likely to result from a 'second opinion' scheme for decisions on biopsy.

Although the cost comparison between different forms of biopsy was reliable there may be methodological problems with the comparison with mammography. It was not clear that the populations concerned were comparable nor was there any discussion of false positives/negatives, or effects of either technique on long-term outcome.

35 Evens, R. G. and Jost, R. G. (1977). The clinical efficacy and cost analysis of cranial computed tomography and the radionuclide brain scan. *Seminars in Nuclear Medicine* **7**(2), 129–36.

1. Study design

1.1. *Study question*

Should cranial computed tomography (CT) be substituted for radionuclide brain scans (RBS) as the first diagnostic procedure in patients with suspected intracranial pathology?(c)

1.2. *Alternatives appraised*

CT *versus* RBS.

1.3. *Comments*

The approach adopted in the study did not lead to a clear identification of changes in resource use or patient management as a result of the different strategies. Therefore, the results of the analysis are not really conclusive.

2. Assessment of costs and benefits

2.1. *Enumeration*

The effectiveness of the diagnostic techniques was considered in terms of the information provided. The eventual effect on patient outcome, through changed patient management, could not be estimated. Possible reductions in the use of other diagnostic tests was discussed.

The costs considered were hospital costs involved in providing the diagnostic services. Although professional fees were also incurred, they were not included in the cost comparison. Costs to the patient and costs arising from resulting changes in patient management were not considered.

2.2. *Measurement*

The study was not based on a randomized controlled trial. Effectiveness was measured in terms of the true positive rate (sensitivity) and true negative rate (specificity) of the two scanning procedures and overall accuracy was determined. Cost information was collected from 98 units providing CT, covering both fixed and variable costs. Costs for RBS were estimated from the experience of the authors' own unit. Although a variety of scans were performed on the same equipment, the costs were

Evens and Jost (1977)

calculated as if a dedicated facility were provided rather than assessing the marginal costs for the existing provision.

2.3. *Explicit valuation*

Data on specificity and sensitivity were taken from the existing literature. Market values were probably used for cost estimation, although no details were given.

3. Allowance for differential timing and uncertainty

Equipment costs were depreciated over a five year period. However, the annuitization procedure was not used. (See Chapter 2.)

The authors presented a range of sensitivity and specificity results from the literature and also discussed the effects of throughput on their average cost estimates.

4. Results and conclusions

The literature surveyed suggested that CT was generally more sensitive than RBS, although there was considerable variation in the results for different units and different conditions. Three studies of specificity showed high performance (97–100 per cent) for both tests. Overall accuracy depended on the range of conditions considered. CT was reported to be between 20 and 30 per cent more accurate if conditions detectable by CT but not RBS were included.

The average cost per CT scan was estimated to be US$130, with a range of US$89–US$157 depending on patient throughput.

The comparable average cost estimate for RBS was US$51. Allowing for the difference in accuracy, the average cost per correct diagnosis was US$141 for CT and US$73 for RBS.

The authors pointed out that the use of CT as the first diagnostic test may reduce the use of other tests. A negative RBS may lead to a CT scan as a second test, whilst the reverse is unlikely. Reductions in the use of other diagnostic tests were also reported in the literature. However, the cost savings implications of this were not calculated.

The authors concluded their argument in favour of CT by suggesting: 'that whether lives or years are saved, a correct diagnosis . . . is necessary to stop the diagnostic workup, begin appropriate therapy and determine prognosis'.

5. General comments

Although the execution of the study could have been improved, it does provide an example of the problems that arise when a procedure is found to be both more costly and more effective. In this case, the possibility of cost saving induced elsewhere in the health care system was not accounted for and this may assist in resolving the difficulty. However, the interpretation of the existing results could be made more clear.

The decision of whether to use a more expensive, but more effective, test should be based on marginal costs and marginal benefits, but marginal costs are not presented in this study. (For an example of this approach see Weinstein and Fineberg 1978, and Hull *et al.* 1981, also reviewed in this section.)

36 Holmin, T., Jönsson, B., Lindgren, B., Olsson, S.-A., Peterson, B. G., Sörbris, R., and Bengmark, S. (1980). Selective or routine intraoperative cholangiography: a cost-effectiveness analysis. *World Journal of Surgery* **4**, 315–22.

1. Study design

1.1 Study question

Is it more cost-effective to perform intraoperative cholangiography selectively or routinely?(c)

1.2. Alternatives appraised

Routine intraoperative cholangiography *versus* selective intraoperative cholangiography (on patients fulfilling at least one of four common clinical criteria). (See 1.3 below.)

1.3. Comments

The authors mentioned that there are methods of investigating the biliary tract and that these methods can be combined in different ways and be linked with different clinical criteria. It was assumed in the study that the only purpose of intraoperative cholangiography was to detect common bile duct stones. The four criteria (of which any one was used to indicate selective cholangiography) were colic attacks, cholecystitis, pancreatitis, and jaundice.

2. Assessment of costs and benefits

2.1. Enumeration

The costs considered were the resource costs of the two procedures, plus any costs arising from positive findings, and the morbidity and mortality associated with the two approaches to patient management. The benefits were assumed to be related to the detection and treatment of stones in the common bile duct.

2.2. Measurement

Clinical data were based on a retrospective review of case notes from 148 patients who underwent common duct exploration and a 25 per cent random sample of 671 who had cholecystectomy alone. Cost data were based on an earlier calculation of hospital costs for gallstone operations.

2.3. Explicit valuation

Market prices (hospital expenditures) were used to estimate costs. No attempt was made to value benefits.

3. Allowance for differential timing and uncertainty

Differential timing was not considered. It was not particularly relevant in the context of this study since all the costs and effects occur within a short space of time.

No sensitivity analysis was performed. Given that many of the clinical and economic estimates were drawn from different sources, a sensitivity analysis would have been beneficial.

4. Results and conclusions

Selective use of intraoperative cholangiography would have led (in the patients considered) to lower costs and lower mortality rates. The main reason for this outcome was that the benefits of detecting more stones in the bile duct by the routine use of cholangiography were counterbalanced by an increase in costs and risks because false positive operative cholangiograms led to a greater number of unnecessary choledochotomies.

The authors also pointed out that it is by no means certain that cholangiography is the ultimate method of detecting stones in the common bile duct and that there is great need for studies comparing other techniques with routine or selective cholangiography.

5. General comments

This study is one of many employing the use of decision trees to set out the sequence of clinical diagnostic and therapeutic options.

37 Hull, R., Hirsh, J., Sackett, D. L., and Stoddart, G. (1981). Cost-effectiveness of clinical diagnosis, venography and noninvasive testing in patients with symptomatic deep-vein thrombosis. *New England Journal of Medicine* **304**, 1561-7.

1. Study design

1.1. *Study question*

What is the most effective and cost-effective method of diagnosing clinically suspected deep-vein thrombosis (DVT)?(c)

1.2. *Alternatives appraised*

Clinical diagnosis *versus* venography *versus* a combination of impedance plethysmography (IPG) and leg scanning. (See 1.3 below.)

1.3. *Comments*

In principle, the study protocol enabled a comparison of a range of diagnostic strategies. All patients underwent clinical diagnosis; those with positive diagnosis underwent IPG. If the result was positive, venography was performed, if negative, leg scanning and repeated IPG were performed. In cases where DVT was confirmed, the patient received anticoagulant therapy. The logic for the study was that more accurate diagnosis could avert potentially costly therapy.

2. Assessment of costs and benefits

2.1. *Enumeration*

A total of 516 patients were studied between 1975 and 1979. The costs of diagnosis included the direct cost of the test plus the associated treatment costs for a positive result. The criterion for effectiveness was correct identification of a case of DVT.

2.2. *Measurement*

The costs of the individual diagnostic tests and a course of anticoagulant therapy were calculated. (The latter included drugs, laboratory tests and hospitalization charges. These were calculated by considering only the 'hotel' element of hospital average costs.)

These cost data were then combined with the clinical data on case finding rates by each test to derive the total costs (and cost per patient) of employing each diagnostic strategy. (Details of the costing methods are given in an appendix to the paper.)

128

The costs of side-effects of diagnosis and therapy were considered but not measured since they were low in comparison to the other costs.

2.3. *Explicit valuation*

The costs used were physician and other third-party charges, plus daily operating costs incurred in an Ontario hospital. The economic analysis was repeated using costs from a New England urban teaching hospital.

3. Allowance for differential timing and uncertainty

Changes in the following factors were examined by sensitivity analysis: the prevalence of DVT, hospitalization costs, costs of anticoagulant therapy, and the costs of the diagnostic tests. In general sensitivity analysis did not alter the ranking of the findings. However, the relatively high cost of venography in the United States *vis-à-vis* Canada did make it seem less economically attractive in the United States.

4. Results and conclusions

Total costs, and the cost per case correctly diagnosed were calculated for varying combinations of diagnostic tests. The cost per patient with DVT correctly identified and treated by clinical diagnosis alone would be Canadian $6319, and for proximal DVT $9137. The cost per case diagnosed using outpatient venography would be $2840 ($4107 for proximal DVT), and using venography as an elective inpatient diagnostic procedure $3908 ($5651 for proximal DVT). The cost per case diagnosed using IPG alone was $2784 ($2995 for proximal DVT), using IPG plus leg scanning $2989 ($3986 for proximal DVT), and using IPG plus outpatient venography $2874 ($4155 for proximal DVT).

Clinical diagnosis was therefore cost-ineffective, venography was more cost-effective, especially on an outpatient basis. IPG was a suitable alternative for patients with DVT, and was less invasive, but not for patients with proximal DVT for whom IPG plus leg scanning was a suitable alternative. Inpatient costs were a major element of all diagnostic and treatment methods, therefore outpatient methods should be considered where possible.

The authors pointed out that the correct way to interpret their results was by *incremental* cost-effectiveness analysis; that is, to consider the *additional* cost per case found by one diagnostic strategy over another. (See the methodological introduction to this volume.)

Hull *et al.* (1981)

5. General comments

Only direct costs of diagnosis and treatment were included. Costs to patients (e.g. time and inconvenience) were not considered. Otherwise, this economic appraisal is above average, with a careful methodology for the clinical and economic aspects and detailed costings.

8 Knaus, W. A., Wagner, D. P., and Davis, D. O. (1981). CT for headache: cost/benefit for subarachnoid hemorrhage. *American Journal of Roentgenology* **136**, 537–42.

1. Study design

1.1. *Study question*

What are the costs and benefits associated with finding a case of subarachnoid haemorrhage via computed tomography (CT) applied in cases of headache?(d)

1.2. *Alternatives appraised*

CT scan of headache patients *versus* CT scan of headache patients *only* if they have other symptoms as well.

1.3. *Comments*

2. Assessment of cost and benefits

2.1. *Enumeration*

Only costs incurred by the hospital, for diagnosis and therapy, were considered.

All requests for scans during the study period were placed into diagnostic categories, and the diagnostic yield (i.e. positive scans per total scans) calculated for each category. Those diagnosed as having subarachnoid haemorrhage were followed for short-term outcome. The study examined the diagnostic paths and clinical outcomes for patients from diagnosis to final outcome.

2.2. *Measurement*

The study was not linked to a controlled clinical trial. Using diagnostic yields and average CT costs, the cost per positive case found was calculated for all categories of patients. From the case finding costs, the national costs were calculated for the two alternative diagnostic strategies being examined. The incremental costs of scanning all headache patients were then calculated. From the incremental costs, and morbidity and mortality figures, the cost per year of life saved was calculated.

2.3. *Explicit valuation*

The study used the 1977 average national CT cost cited in the literature

Knaus et al. (1981)

(US$255), plus the 1977 average national charges for diagnostic alternatives, the costs for high risk operations, and costs of hospitalization.

It was not stated how these costs were derived, it is assumed they reflect market prices.

3. Allowance for differential timing and uncertainty

Discounting was not applied. It would be relevant in considering the capital costs of the CT scanner, unless it were argued that the charges for CT scanning incorporate a capital element. (See Chapter 2.)

Some sensitivity analysis was applied, considering different incidence (case finding) rates. Also, the possibility of false positives and false negatives was considered, and the cost implications explored.

The impact of different patient throughputs on costs was not considered.

4. Results and conclusions

The cost of finding a case of subarachnoid haemorrhage was US$32 895 for all headache patients, and US$1050 for patients presenting with additional symptoms. The incremental cost of scanning all headache patients was calculated to be US$116 million per annum nationally (US$135 million *versus* US$19 million). Performing sensitivity analysis with respect to incidence rates gave total costs lying between US$239 million (for low incidence rates) and US$12 million (high incidence rates).

Applying different incidence rates and using the net costs of scanning and surgery (i.e. adjusted gross costs), the cost per life saved was calculated as lying between US$1.1 million (low incidence rates) and US$44 000 (high incidence rates), and the cost per life year saved between US$50 000 (low incidence rates) and US$2000 (high incidence rates).

Taking false negatives into account, if the rate were 10 per cent the cost per life saved would be increased by about 11 per cent due to the increase in number of deaths owing to diagnostic delay. Taking false positives into account, if the rate were 10 per cent the cost per life saved would again be increased by about 11 per cent due to the costs of unnecessary tests and extra hospitalization.

The costs per life year saved were compared with similar costs from other studies.

5. General comments

The authors considered that the technique employed should serve as a model for this type of analysis. Although this way of appraising diagnostic tests is of interest, this particular study would have been greatly improved by more careful costing methodology.

39 Larson, E. B., Omenn, G. S., and Lewis, H. (1980). Diagnostic evaluation of headache. Impact of computerised tomography and cost-effectiveness. *Journal of the American Medical Association* **243**(4), 359–62.

1. Study design

1.1. *Study question*

What is the impact of computed tomography (CT) on the care of outpatients presenting with headache?(a)

1.2. *Alternatives appraised*

CT *versus* other diagnosed procedures.

1.3. *Comments*

Three cohorts were formed retrospectively, one pre-CT, the second immediately after CT, and the third one year after CT. Statistical techniques were applied to measure the prevalence of abnormal findings in the population from whom the patients were taken. The study was not linked to a controlled clinical trial.

2. Assessment of costs and benefits

2.1. *Enumeration*

The study used the technique of cost-effectiveness and hence assumed the diagnoses across the methods were similar, hence benefits were not considered directly. Only the costs of the diagnostic procedures were included; the costs of any subsequent therapy were not considered. The study considered hospital costs only; patients' costs were not considered.

2.2. *Measurement*

The total charges accruing to each cohort were calculated and then the average cost per patient for each cohort, by diagnostic procedure. The total number of abnormal findings was used to calculate the minimum cost (with 95 per cent certainty) of finding an abnormal case.

2.3. *Explicit valuation*

All costs were standardized to the 1976 billing schedule, and hence it is assumed they represented market values.

3. Allowance for differential timing and uncertainty

No sensitivity analysis was performed. Although descriptions of the patient caseload and the number of procedures used were given, the authors did not examine the effect of changing patient numbers, or incidence rates, and there was no analysis of different test combinations. Discounting was not explicitly applied to capital costs. (See Chapter 2.)

4. Results and conclusions

During the study period it was found that CT replaced other investigations but not sufficiently to offset the extra costs of CT scanning.

The total charges to patients before CT were US$10 252, immediately after CT they were US$8952 and one year later they were US$11 641 (CT accounting for 50 per cent). The average charges per patient were US$183.07, US$179 04 and US$211.65 respectively. The cost of finding a case of brain tumour was estimated to be at least US$1265 for patients with abnormalities on neurological examination and US$11 901 for patients with normal findings on neurological examination.

The authors stated that careful examination was needed to identify patients who required CT scans, as opposed to other diagnostic procedures. They considered that skull roentgenograms were of no diagnostic value, and that the routine use of any procedure was likely to increase the number of false positives, and hence the cost of a true positive.

5. General comments

The authors felt they may have had a highly selected group and hence the results may not be generally applied. Also the costing methodology was weak, since the authors did not say how the costs were calculated or derived. However, our main reservation is that the study considered only one aspect of the economics of CT scanning. Other aspects include the impact of scanning on therapy, the less invasive nature of CT, the reassurance value of true negatives and the impact of CT on patient outcome.

135

40 McNeil, B. J., Thompson, M., and Adelstein, S. J. (1980). Cost-effectiveness calculations for the diagnosis and treatment of tuberculosis meningitis. *European Journal of Nuclear Medicine* **5**, 271–6.

1. Study design

1.1. *Study question*

What is the most cost-effective combination of new and existing diagnostic tests and therapy for tuberculosis meningitis (TBM)?(c). (See 1.3 below.)

1.2. *Alternatives appraised*

Combinations of three diagnostic techniques (clinical diagnosis, radio-bromide partition test, or a theoretically perfect test) and two therapies (conventional treatment with para-aminosalicyclic acid (PAS) and isonicotinic acid hydrazide (INH) and streptomycin, or 'second line' therapy with rifampin or ethambutol).

1.3. *Comments*

The setting for the study was a developing country, India. A decision tree was utilized, following a population from presentation of TBM through diagnosis, treatment and hospital care to outcome. Probabilities of mortality and cure were applied to the costs associated with the treatments and the four possible diagnoses (i.e. true positive, false positive, true negative, and false negative). The authors were particularly interested in exploring the trade-offs between investments in more accurate testing and more effective therapy, given that countries with limited resources may not be able to institute both.

2. Assessment of costs and benefits

2.1. *Enumeration*

The costs included were those for inpatient and outpatient treatment, drug therapy for given lengths of time, and diagnostic tests. Benefits were quantified in the form of health results, e.g. number of cures, and number of preventable deaths. No patients' costs were included, nor loss of productive time, although these may be difficult to quantify in a developing country.

136

2.2. *Measurement*

Calculations of costs and outcomes were based on figures from the existing literature. Two ranges of costs were calculated for the diagnostic tests (an upper and lower). Five categories of costs were calculated at two disease prevalences: (1) total costs for diagnosis using the two cost schedules, (2) total cost for treatment using both therapies, (3) total costs for diagnosis and treatment using the lower cost schedule, (4) the average cost of case finding using both cost schedules, and (5) the marginal costs of achieving an additional cure using either a new diagnostic test instead of the current method, or 'second line' therapy along with existing diagnostic techniques.

2.3. *Explicit valuation*

All financial costs were taken from estimates in other studies. The authors did not say how the costs were calculated but probably market values were applied. All costs were given in Indian rupees (R) at an unstated price level.

3. Allowance for differential timing and uncertainty

Two prevalence rates for TBM were considered, 30 and 80 per cent. Also, two schedules of costs were calculated for diagnostic tests, and a range of average costs of case finding presented.

Discounting was not applied, although it was not particularly relevant as all costs occurred within a short space of time.

4. Results and conclusions

The average cost per case found at 30 per cent prevalence was R224 with clinical diagnosis, between R211 and R248 using the bromide diagnosis, and between R207 and R257 using the perfect test. At 80 per cent prevalence, the average cost per case found fell to R84, R79–R93, and R78–R96 respectively. Direct comparisons were also made across the diagnostic techniques (including number of cures, number of hospital days, and total costs of diagnosis and therapy) if combined with conventional or 'second line' therapy.

The total costs of diagnosis and therapy for 1000 patients were as follows. At 30 per cent prevalence, the costs varied from R1.084 × 10^6 using conventional therapy and treatment to R1.804 × 10^6 using the perfect test and second line therapy. At 80 per cent prevalence, the costs varied from R1.352 × 10^6 to R2.072 × 10^6 respectively.

McNeil *et al.* (1980)

Marginal costs for achieving an additional cure using either new diagnostic techniques or second line therapy were calculated. At low prevalence rates of TBM, they showed that there was a net saving from the introduction of either of the new tests. At high prevalence rates, total costs of therapy rose, resulting in marginal costs of between R694 and R958 per additional cure. The introduction of second line treatment led to a significant cost increase at both prevalence rates.

5. General comments

The methodology employed in this study was good and could be applied to other situations. Whilst the range of costs calculated was thorough, the costs were not original to the study and the authors do not indicate how accurate they are.

41 Neuhauser, D. (1978). Cost-effective clinical decision making: are routine paediatric pre-operative chest X-rays worth it? *Annales de Radiologie* **21**, 80–3.

1. Study design

1.1. *Study question*

Are routine paediatric pre-operative chest X-rays worthwhile?(d) (See 1.3 below.)

1.2. *Alternatives appraised*

Routine paediatric pre-operative chest X-rays (including follow-up or modification of therapy) *versus* no X-rays.

1.3. *Comments*

Although the study question is as cited, a major objective in the paper was to illustrate how decision analysis can be used to structure and to analyse a clinical decision making problem. Therefore some of the data were merely illustrative and the author did not seek to justify, nor to check, their accuracy.

2. Assessment of costs and benefits

2.1. *Enumeration*

The costs considered were the resource costs of X-rays, follow-up tests, and therapy modification resulting from positive X-rays (true and false), including prolongation of hospital stay. In addition, the costs, in terms of mortality and morbidity, of radiation exposure were mentioned but not measured. The benefits identified were the change (for the better) in surgical practice based on X-ray findings and the discovery of unsuspected abnormalities.

2.2. *Measurement*

The details of cost measurement were not given since these were taken from another study. The benefits identified above were converted, with the help of a few assumptions, to expected life years gained.

The frequency of changes in surgical practice and discovery of unsuspected abnormalities were estimated from an analysis of 1500 consecutive X-rays performed on children aged 0–19 years.

139

Neuhauser (1978)

2.3. *Explicit valuation*

Market prices, probably hospital and physician billings, were used to estimate costs. Life years gained were not valued, but the implied cost per life year gained by this procedure was compared to that gained by others, such as 'increased polio immunization or urging the surgical residents to use a stethoscope'.

3. Allowance for differential timing and uncertainty

Life years occurring in the future were discounted to present values at a rate of 4 per cent. It was tacitly assumed that all costs occur in the current year.

The usefulness of sensitivity analysis, given the number of assumptions made, was discussed; but sensitivity analysis was not employed. The main assumption, upon which an analysis could be performed, was that the change in surgical practice results in a 5 per cent reduction in operative mortality.

4. Results and conclusions

Given the assumptions made in the article, the cost per present value year of life gained by this procedure was around US$103 000. This should be compared with the costs implied by other options. The author claimed that the comparable figure for (say) expanding polio immunization was only US$15 000. In order to yield benefits of around US$15 000 per present value life year gained, this procedure would have to result in 4.6 present years of life saved (on average) owing to the earlier discovery of unsuspected abnormalities. The author also discussed the possibility of performing pre-operative X-rays on a sub-group of patients, rather than all.

5. General comments

This article, being very short, contains a concise discussion of the economic trade-offs implicit in clinical decision making. One of the main objectives was to follow up a statement made in a medical paper on the same topic, namely that the X-rays were 'medically and economically justified and essential'.

42 Royal College of Radiologists (1981). Costs and benefits of skull radiography for head injury. *Lancet* **ii**, 791–5.

1. Study design

1.1. *Study question*

What are the costs and benefits of routine skull radiography for patients admitted to Accident and Emergency Units with head injury?(d)

1.2. *Alternatives appraised*

Routine skull radiography *versus* clinical diagnosis unaided by radiological evidence, in the management of head injury patients.

1.3. *Comments*

In analysing the costs and benefits of routine skull radiography, the authors differentiated between those for patients presenting with complicated and uncomplicated head injury. It was noted that no radiography *versus* routine radiography comprised the extremes of the options available to clinicians and this paper was therefore intended as a framework for further discussion.

2. Assessment of costs and benefits

2.1. *Enumeration*

Costs and benefits were considered from the perspective of the NHS.

In costing the alternatives, the authors took account of radiography costs and the effects of radiography use on admissions policy and consequently upon inpatient costs.

The benefit measure was the number of intracranial haematomas correctly diagnosed.

2.2. *Measurement*

The study was conducted as part of a prospective study of nine Accident and Emergency Units in England, Scotland and Wales, and included 5850 patients who received skull radiography (i.e. approximately 1 per cent of the annual national case load).

Within the study, each doctor requesting skull radiography was required to record the presence of any clinical signs of underlying brain damage, the disposal of patients following radiography and the likely disposal of patients in the absence of skull radiography. Whilst the length

141

of stay figures were drawn from the study, the inpatient and radiography costs used were averages drawn from the literature.

2.3. *Explicit valuation*

Market values were used to estimate health service costs. No attempt was made to assign a value to the successful early detection of intracranial haematomas.

3. Allowance for differential timing and uncertainty

These were not explicitly considered.

Given the low incidence of intracranial haematomas, the authors acknowledged the difficulty of attempting to mount a sufficiently large study to calculate precisely the contribution of radiography in this clinical application.

4. Results and conclusions

Of the 5850 patients admitted to the study 1021 had complicated head injuries. The four intracranial haematomas occurring to patients in this group would have been detected without the aid of radiography on the basis of underlying clinical evidence. Of the 4829 patients with uncomplicated head injury, four had intracranial haematomas, but only three of these would have been diagnosed in the absence of routine radiography.

Comparing stated admission policy in the absence of radiography with actual admission policy following radiography, 70 per cent more patients with uncomplicated head injury would have been admitted in the absence of radiography.

Of those patients admitted, the average length of stay with radiography was 1.3 days and without radiography would have been 1.4 days.

Taking account of the resulting radiography and inpatient costs, it was demonstrated that the use of radiography to detect the single case of intracranial haematoma which was clinically negative resulted in additional health service expenditure of £10 954. Thus, the radiological cost of £42 849 would be partly offset by the savings in hospitalization, as more patients would be admitted in the absence of radiography. Extrapolating from the results of the study, the authors suggested that the use of routine skull radiography would increase rather than decrease NHS expenditures. In the context of comparative risk, for patients with uncomplicated head injuries, the risk of having an undetected intracranial haematoma in the absence of skull radiography would be no greater than that of a male of any age being killed in a motor vehicle accident.

5. General comments

It would have been appropriate for sensitivity analysis to have been applied to demonstrate the effects of differences in incidence, admission policy and length of stay upon the results generated.

As the authors acknowledged, use of average costs in the case of radiography costing would probably result in overestimation of radiography cost savings as a result of discontinuation of routine skull radiography; as would the use of an average cost per inpatient day overstate the effects of different policies' costs in terms of use of inpatient resources.

The study could have been developed through consideration of the benefits resulting from early detection of intracranial haematomas and assigning a monetary value to these to permit a full cost–benefit analysis.

43 Weinstein, M. C. and Fineberg, H. V. (1978). Cost-effectiveness analysis for medical practices: appropriate laboratory utilization. In *Logic and economics of clinical laboratory use* (E. S. Benson and M. Rubin, eds). Elsevier/North-Holland Biomedical Press, Amsterdam.

1. Study design

1.1. *Study question*

How cost-effective are alternative screening strategies for diagnosing pheochromocytoma in hypertensive patients?(c)

1.2. *Alternatives appraised*

Universal screening *versus* screening patients with symptoms in addition to high blood pressure *versus* treating hypertension for one year then screening patients with uncontrolled blood pressure *versus* screening symptomatic patients now and screening asymptomatic patients if blood pressure is uncontrolled after one year *versus* no screening.

1.3. *Comments*

Further specific screening strategies could be devised but those selected by the authors cover the range of possibilities in a reasonable way. The logic was that attention could be paid to refinements of the strategies when the initial results were known.

2. Assessment of costs and benefits

2.1. *Enumeration*

The only costs considered were health care resource costs. Other costs falling on the patient or elsewhere in the economy were ignored. Health care costs comprised the cost of testing for excessive vanylmandelic acid (VMA) and net health care costs induced as a result of the test information, the change in disease morbidity, side-effects of tests and treatments and the additional life years gained. Net benefits were expressed as quality-adjusted life years gained.

2.2. *Measurement*

Illustrative costs and benefits were calculated for a particular cohort of patients; 30-year-old men with diastolic blood pressure of 110 mg. Typical treatment costs and factors such as disease prevalence, risk of death, side effects, and the sensitivity and specificity of the screening test were taken from existing literature.

2.3. *Explicit valuation*

Details of the valuation methods used were not given but it is likely that the normal scales of fees and charges for health care services were used.

3. Allowance for differential timing and uncertainty

All costs and benefits were discounted at 5 per cent per annum.

The importance of sensitivity analysis was stressed and two examples were set out. These related to the prevalence of pheochromocytoma and the diagnostic sensitivity and specificity of the symptomatology used to prescreen in some of the strategies.

4. Results and conclusions

The results were given for each strategy, under the different assumptions, and presented in terms of the cost per quality-adjusted life year saved relative to the no screening option. The authors gave detailed attention to the interpretation of the results for decision making purposes, stressing the relevance of both the efficiency of the screening strategy and the decision maker's valuation of life years saved. The decision points between the different strategies were shown by means of graphs. The results ranged from US$420 000 per life year saved under universal screening with low disease prevalence, to US$7900 for screening only symptomatic patients, assuming high disease prevalence.

5. General comments

The particular example of screening for pheochromocytoma is part of a longer article on the cost-effectiveness of laboratory expenditures. There is a full discussion of the cost-effectiveness technique and the particular issues that arise in the use of laboratory tests; for example, different concepts of effectiveness and the problem of allocating joint costs.

SECTION 4: Alternatives in therapy

Introduction

The majority of health care resources are used in therapeutic interventions of one sort or another, so it is fitting that this section is the most extensive in the volume. There are 21 studies, five dealing with alternative treatments for chronic renal failure, four from the field of cancer therapy, two concerned with drug therapy for duodenal ulcer, and a number of others from a wide range of medical specialities including rheumatology, obstetrics, psychiatry (alcoholism), paediatrics, ophthalmology, and surgery. There is also one study originating from a developing country, even though the majority of evaluations in the developing world are understandably concerned with preventive programmes and were summarized in Section 2.

Unlike the studies reviewed in the section on diagnostic alternatives, the studies summarized here are not exclusively concerned with 'high technology', although they include evaluations of continuous ambulatory peritoneal dialysis, bone marrow transplants, neonatal intensive care, and cimetidine, an H_2 receptor antagonist which it is suggested may avert, or delay, the need for surgery in duodenal ulcer disease. Nevertheless, there are also evaluations of 'low level' technologies, such as methods of reducing complications following surgery and of counselling following mastectomy.

Particular methodological problems in this area

The major problem facing those wishing to evaluate alternatives in therapy is that of securing appropriate medical evidence upon which to base their economic analysis. It was pointed out in the methodological introduction to this volume (Chapter 2) that the proportion of economic appraisals carried out in association with clinical trials is growing, although it still remains small in relation to the total. Furthermore, it was stressed that where trials have been carried out they may be of poor quality, e.g. insufficient numbers of patients in the trial to identify a clinically significant difference at the 5 per cent level (of statistical significance) if indeed such a difference exists.

The problems in linking economic analysis with clinical trials are best illustrated by the study by Culyer and Maynard (1981), summarized in this section. They point out that despite the large number of trials of cimetidine, the new drug they wished to evaluate from an economic viewpoint, none provided a suitable foundation for the economic analysis. The existing trials either evaluated inappropriate alternatives (such as a placebo which was unlikely to be administered in actual medical practice), were not well controlled, were too small, or included an inadequate range of measurements. For example, measurements in trials are typically confined to a fairly narrow assessment of medical 'improvement' or 'cure' and do not include the broader assessments of the patient's functioning (or that of his family); these would be of importance in a subsequent economic analysis, such as the ability of the patient, or a member of the family, to return to work.

Given the high level of investment in medical research in most developed countries, and the concerns about the costs of new medical technologies, it is surprising that no coherent strategy has emerged for incorporating economic analysis into clinical trials, i.e. *when* should economic analysis be included in a given trial, and *how* should that analysis be carried out? These issues have been addressed by Drummond and Stoddart (1984) who argue that whilst it would be unnecessary to include economic analysis in *every* trial—especially those of the 'efficacy' type*—it is possible to identify situations where economic analysis would be indicated. These include situations where there are imminent resource allocation decisions concerning the diffusion of the new therapy, where there would be large resource consequences (either because the new therapy would be given to a large patient population or because it is radically different from existing therapies), and where resource considerations are prominent in the decision to set up the trial. Those involved in medical research should consider whether, on the basis of these criteria, any of their own trials should incorporate economic analysis. Drummond and Stoddart (1984) also give advice on the *phasing* of the economic analysis so as to minimize wasted time and effort. This is important since the results of the *clinical* analysis may not be known

---

* 'Efficacy' or 'explanatory' trials are primarily concerned with establishing whether a therapy has a biological effect when delivered under *ideal conditions*. They are to be contrasted with 'effectiveness', 'pragmatic' or 'management' trials, which are concerned with assessing the therapy's impact when delivered in the context of prevailing health care organization and practice. There are a number of crucial differences in the design of the two types of trial, since they have quite different purposes. For more details see Drummond and Stoddart (1984).

when the economic analysis is being planned. The precise form of the economic analysis (e.g. whether or not cost-effectiveness will suffice, or whether cost–benefit analysis is required) may depend on the answer to the clinical question. Broadly speaking the strategy suggested by Drummond and Stoddart is to collect those data during the trial that are patient-related and hence difficult or expensive to extract later, e.g. patient and family costs, length of hospital inpatient stay, utilization of major resource items such as medical time, nursing time, and diagnostic tests.

We believe that the more frequent inclusion of economists in multidisciplinary teams undertaking research into new medical therapies will yield other important benefits. First, it will improve the quality of the economic analysis typically observed in such studies. The methodological comments made on the studies illustrate that on occasions authors would have benefited from economic advice. Secondly, involvement in multidisciplinary work may help economists make their work more relevant, particularly in the selection of alternatives for appraisal (e.g. in linking these more closely to choices in clinical policy or health care planning) and in the assessment of the benefits of interventions (e.g. developing quality of life measures).*

The other major methodological problem facing those wishing to analyse choices in therapy is that of estimating hospitalization costs. Some comments on this problem were made in Chapter 2 and these will not be repeated here. However, there is further discussion of the current state of the art in the estimation of hospitalization costs below.

Current state of the art

Two features of the current literature will be highlighted here; the assessment of the benefits of therapies in quality-adjusted life years and the estimation of hospitalization costs.

Turning first to assessment of benefits in quality-adjusted life years, it was mentioned in the methodological introduction to this volume that there are two distinct approaches: (*i*) the *measurement* of preferences for health states by the relevant population, and (*ii*) the estimation of utility values by a quick (and inexpensive) *consensus* forming exercise, followed

---

* A third area where economists may have a useful contribution to make is in assessing the value for money from medical research itself. Although outside the scope of this volume, there are a number of interesting issues surrounding the development of medical research priorities, the development of more cost-effective research protocols taking advantage of economies of scale, and the production of more timely and more relevant results. These issues have already interested a number of authors. See Black and Pole (1975) and Hawkins (1984).

by sensitivity analysis. In the studies in this section, Boyle *et al.* (1983) use the former approach, whereas Weinstein (1980) uses the latter. Expression of the outcome of therapy in quality-adjusted life years is useful where the clinical alternatives produce different results in terms of mortality and morbidity, and where one wishes to have a common unit of outcome that combines both effects. (It is also useful where the *quality*, as opposed to quantity, of life extension is the crucial outcome of interest.)

One would have expected more of the studies to have explored the quality of therapeutic intervention, at least in its more limited form as used by Weinstein (1980), since many of them compare surgical treatment with a medical alternative. In this case one would expect the treatment options to produce complications of a different nature and duration. For example, what are the pros and cons, in treatment for duodenal ulcer, of a surgical operation with its attendant risks or a lifetime drug regimen? In treatment for chronic renal failure, what are the relative valuations of a year of life extension gained by transplant, hospital dialysis, home dialysis and continuous ambulatory peritoneal dialysis? Who should make these judgments? Clearly evaluations of these alternatives should address these issues to some degree, although that may not necessarily imply measurements of quality of outcome. It may be, for example, that the option presumed to have higher quality of outcome also dominates in terms of cost. (This is often found to be the case in transplantation for chronic renal failure.) However, even in this case one ought to check whether all people have an ordinal preference for transplant; some very risk-averse individuals may not wish to have an operation. The area of chronic renal failure is one where researchers have tried to measure patients' preferences. (See Churchill *et al.* 1984, and Sackett and Torrance 1978.) However, this work is not always reflected in the studies summarized here.

In situations where the costs of the alternative treatments are similar and where it is agreed that patients' assessments are the most relevant ones, it may not be necessary to *measure* patients' preferences, but merely offer them the choice of therapy (after appropriate explanation of the expected outcomes). However, the situation may be complicated by the fact that the costs per patient of delivering therapy are related to the size of provision in a given locality. For example, there are likely to be economies of scale in the provision of facilities such as minimum care dialysis units.

There is now much more experience in the measurement of individuals' preferences for health states than existed when the first volume of *Studies in economic appraisal in health care* was published. There has also been more exploration of the preferences of different groups in the community. In the earlier work most of the respondents were physicians, presumably because they were readily available and already understood the implications of being in given states of health. In the study by Boyle *et al.* (1983), measurements were made of the preference of a random sample of parents in the city where the neonatal unit was based. The logic was that this would be the group whose preferences were most relevant to the *health care planning* decision to institute a regional neonatal programme. In other situations, such as the choice between therapies of similar cost, it may be decided that the patient's preference is the relevant one.

The measurement of preferences for health states of patients or the general public raises a number of practical and ethical issues. Much progress has been made with the practical issues, such as the reliability and reproducibility of measurements, and a number of measurement instruments have been developed. (For more details see Drummond *et al.* Chapter 6, 1985.) There is less unanimity regarding the ethical issues, such as the possibility that individuals (including patients) could be placed in a stressful situation for the purposes of research.

The other developments worth commenting on here are in the estimation of the costs of hospitalization. Whereas the majority of studies still use average hospital (or *per diem*) costs, albeit with varying degrees of scepticism and qualification, some use more sophisticated methods. For example, Boyle *et al.* (1983) develop a simultaneous allocation method for the apportionment of hospital costs to individual patient care departments. (Their particular interest was the neonatal intensive care unit.) The simultaneous approach adjusts for the fact that not only do the hospital overhead department service the patient care departments, but also service each other. For example, the domestic services department cleans the administration block and itself requires the services of the administration department, in the recruitment and remuneration of staff. The simultaneous approach requires that units of departmental output be developed and assessments made of the level of service given to each department by all of the others. A significant amount of work is involved, but the approach may be justified if the results of a given study are highly sensitive to the estimates of hospitalization costs.

Of course the general approach recommended by economists is that of marginal analysis. That is, to ascertain which (if any) of the hospital costs would change if a given treatment programme were added to, or subtracted from, the overall activity. Often little will change if one programme is added or subtracted. Whilst this is fine up to a point, the most common situation is that the choice does not concern such an addition or subtraction, but is between two programmes, one of which may free more central services for alternative uses. In this case one does require an estimate of the level of services freed and their value in other treatment programmes.

Contribution

In examining the contribution to decision making of the studies summarized, it is important to recognize that there are two levels (or types) of decisions such studies could inform. First, there are *health care planning* decisions about whether a given type of service should be made available, or whether a new medical technology should be allowed to diffuse more widely throughout the health service. Secondly, there are *clinical policy* decisions about the kinds of therapy that should be administered to a patient exhibiting particular clinical signs. If economic evaluations of therapeutic options are to have any impact they should probably influence both sets of decisions, although the mechanisms by which they can influence the latter set are far from clear, given the concern many practitioners have about infringements on 'clinical freedom'.* One of the difficulties in interpreting the results of many of the studies is that the authors do not often explore the implications, either for health care planning or for clinical policy, of their findings. The confusion is often increased by the fact that many of the evaluations are by clinicians of their own practices. Do they intend that the results will influence *their own* practice in the future, or are the results mainly to be used by health care planners and policy makers in making investments in similar therapies *elsewhere*?

The difficulties in influencing the type of therapy given, particularly through changes in clinical policy, are not unique to economic evaluation. It is well known, for example, that clinical trials showing superior effectiveness of one therapy over another, or showing that current therapy is no better than doing nothing, do not always have an impact on physician awareness and clinical actions (Hawkins 1984).

Whilst acknowledging the difficulties, we feel that economic evaluations of therapeutic alternatives are likely to have more impact if they:
—make it clear what kinds of decisions, in health care planning or clinical policy, the results can inform;

* Of course, most economists would argue that clinical actions have never been completely unconstrained, since they have always had to take note of limitations on the availability of resources. However, a full debate on this point is beyond the scope of this volume. Those interested should consult the companion volume, Appendix 1.

—provide good descriptions of the clinical caseload in the unit where the evaluation took place and provide *detailed* descriptions of the therapies delivered. (Some of this information may need to be provided in a supplemetary paper.);

—list the resources consumed by the clinical alternatives in terms of their physical amounts, e.g. hours of medical time, days of hospital stay, as well as giving the costs. (This should assist interpretation of the results in other settings.);

—examine clinical options more frequently in terms of the clinical *indications* for therapeutic intervention;

—produce results in a timely fashion, particularly in relation to the evidence on effectiveness of therapy being obtained from clinical trials. In many cases a closer link should be forged between the economic evaluation and the clinical trials themselves.

References

Black, D. A. K. and Pole, J. D. (1975). Priorities in biomedical research: indices of burden. *British Journal of Preventive and Social Medicine* **29**(4), 222–7.

Churchill, D. N., Morgan, J. and Torrance, G. W. (1984). Quality of life in end-stage renal disease. *Peritoneal Dialysis Bulletin* **4**, 20–3.

Drummond, M. F. and Stoddart, G. L. (1984). Economic analysis and clinical trials. *Controlled Clinical Trials* **5**, 115–28.

——, —— and Torrance, G. W. (In press) *Methods for economic evaluation of health care programmes*. Oxford University Press.

Hawkins, B. S. Evaluating the benefit of clinical trials to future patients. *Controlled Clinical Trials* **5**, 13–32.

Sackett, D. L. and Torrance, G. W. (1978). The utility of different health states as perceived by the general public. *Journal of Chronic Diseases* **31**(11), 697–704.

4 Beardsworth, S. F. and Goldsmith, H. J. (1982) Continuous ambulatory peritoneal dialysis (CAPD) on Merseyside—cost/ cost-effectiveness. *Health Trends* **14**, 89–92.

1. Study design

1.1. *Study question*

What is the relative cost-effectiveness of continuous ambulatory peritoneal dialysis (CAPD) as a form of maintenance dialysis?(c)

1.2. *Alternatives appraised*

CAPD *versus* hospital haemodialysis *versus* minimum care units *versus* home haemodialysis.

1.3. *Comments*

Although it was the authors' intention to establish the relative cost-effectiveness of CAPD *vis-à-vis* other forms of maintenance dialysis, the paper reported what was essentially a costing exercise for CAPD only.

2. Assessment of costs and benefits

2.1. *Enumeration*

For CAPD the costs covered initial training, maintenance and re-admissions. The authors identified the relevant costs as being staff salaries, hospitalization costs, capital expenditure, consumables, patient transport, and social security. The fact that they could not obtain data for the last of these was just as well, since social security payments are, of course, transfers and not resource costs. (See the companion volume, Section 2.1.1.)

2.2. *Measurement*

There were a number of problems with the way in which costs were measured. The authors estimated total annual expenditure for the CAPD programme for each of three years and divided by patient years of treatment to arrive at the cost per patient year for these three years. Not only were the treatment costs themselves average rather than marginal costs, but the costs were averaged over all patients regardless of their stage of treatment. Therefore, after the first year estimates of training costs were averaged over both new and established patients. The same procedure appeared to have been adopted for capital. The costs for re-admission with complications were similarly averaged. No data on the incidence of complications over time were presented; therefore, it was

153

unclear whether this average cost was a reasonable estimate to present. The patients accepted for CAPD were not matched in any way with patients on other therapies. They were on average older and in a poorer medical condition and this would have to be taken into account when comparing costs and outcomes with other programmes.

2.3. *Explicit valuation*

Resources were valued at their cost to the health service. There was some suggestion, however, that dialysis fluid prices were held down by the manufacturers in the early years of the programme.

3. Allowance for differential timing and uncertainty

Discounting was not required as only single year costs were examined and there were no large items of capital expenditure. The distribution of patient costs was not available because individual patient costing had not been undertaken.

4. Results and conclusions

The authors reported that the costs per patient year were £11 916 in 1979, £7221 in 1980 and £5510 in 1981 (all in current prices). The falling costs were partly due to increased numbers being treated with some of the costs fixed and partly to a reduction in the average number of days in hospital. The 'reduction' in training costs was difficult to interpret because of the averaging process adopted. Similarly, the capital costs per patient year were hard to follow.

The authors did not draw any explicit cost-effectiveness comparisons with other forms of renal replacement therapy. They did quote some comparative cost figures but without citing any sources.

5. General comments

This study does not really live up to its title. Even at the level of a costing exercise, however, it is not particularly well executed. Apart from the strange methodology adopted, little account was taken of the lifetime treatment costs for patients starting on CAPD. The authors figures show that of 32 patients followed up for more than one year, half had experienced at least one change of treatment or had died.

The authors mentioned that the price of dialysis fluid had been held down for the first two years of the programme. This highlights a general problem in evaluating new programmes, which is that initial costs to the NHS may not be a good guide to longer-term costs.

15 Bernth-Petersen, P. (1982). Outcome of cataract surgery. IV. Socio-economic aspects. *Acta Ophthalmologica* **60**, 461–8.

1. Study design

1.1. *Study question*

What are the costs and benefits of cataract surgery?(d)

1.2. *Alternatives appraised*

Cataract surgery (monaphakic and biaphakia patients) *versus* no surgery.

1.3. *Comments*

2. Assessments of costs and benefits

2.1. *Enumeration*

The costs considered were the direct costs of a cataract operation. These comprised surgery (including stay in hospital), after-care, medicine, and optics. In addition, the costs of complications and the costs of a second cataract operation (on 40 per cent of patients) were included.

2.2. *Measurement*

The outcome assessment was based on data collected prospectively in a group of 123 consecutive cataract patients. Before surgery, and at one year follow-up, interview data on occupational and socio-economic aspects were collected; there was no control group for ethical reasons. Patients' feelings about the likely deterioriation leading to nursing home care were checked by applying criteria of visual functioning and the likelihood of help from spouse or friends. Economic data related to the 73 monaphakic patients only.

2.3. *Explicit valuation*

Market prices were used to estimate costs and benefits. No attempt was made to value the benefits, in health terms, of cataract operations.

3. Allowance for differential timing and uncertainty

No discounting or sensitivity analysis was employed. These omissions reduce the confidence one could place in the results. The number of patients for whom cataract operation potentially led to a change in occupational or accommodation status was small and therefore the results were quite sensitive to this parameter. Also, the savings from nursing home places avoided spread on average nine years into the future. Discounting by a positive rate would have reduced the size of this term in the cost–benefit equation.

4. Results and conclusions

The benefits (so defined) exceeded the costs. The main reason for public benefits of cataract surgery was the savings on geriatric nursing care. It was argued that, on the basis of the findings in this study, the effect of the 3000 first eye cataract extractions done in Denmark in 1980 was to bring about a saving of around 400 places in nursing homes.

However, the author acknowledged that 'the economic calculations presented do not pretend to be a rigorous cost–benefit analysis'.

5. General comments

This paper was one of a series. The other papers give more details of the clinical aspects of the study.

The paper has a number of weaknesses, which the author acknowledged. In particular, discounting should have been employed. In addition, the savings in disability pensions are transfer payments (see the companion volume, Chapter 2) although these mirror savings in productive output. Also, on the conservative side, a number of the health benefits of cataract operation were not valued.

6 Boyle, M. H., Torrance, G. W., Sinclair, J. C., and Horwood, S. P. (1983). Economic evaluation of neonatal intensive care of very low birth-weight infants. *New England Journal of Medicine* **308**, 1330–7.

1. Study design

1.1. *Study question*

What are the costs and benefits of neonatal intensive care for very low birth-weight infants?(d)

1.2. *Alternatives appraised*

Neonatal intensive care in a health region (including a tertiary referral centre) *versus* no intensive care (for 500–999 g and 1000–1499 g babies).

1.3. *Comments*

2. Assessment of costs and benefits

Costs considered included costs of neonatal intensive care to discharge, follow-up costs (including institutional and special services, appliances and additional/special education) and the costs of parental care at home. In addition the productive output of the two cohorts was calculated.

The effects considered were the differences in mortality and morbidity with and without neonatal intensive care.

2.2. *Measurement*

Data on effectiveness were based on all infants of 500–1499 g birth-weight live born to residents of Hamilton-Wentworth County during the periods July 1964–December 1969 (373 infants) and January 1973–December 1977 (265 infants). These two periods were before and after the advent of neonatal intensive care. Mortality was determined at final discharge from newborn hospitalization. Among children discharged alive from hospital, 121 of 150 survivors from 1964–69 (81 per cent) and 134 of 151 survivors from 1973–77 (89 per cent) were located and surveyed by mail questionnaire.

2.3. *Explicit valuation*

Market prices were used to estimate public agency costs (e.g. hospital and institutional care expenditures) and production losses (e.g. income). An opportunity cost was imputed for provision of parental care at home.

157

Boyle *et al.* (1983)

A classification of health states was developed to measure health of survivors, according to their physical, role and social and emotional function, and health problems. This classification (of 960 possible health states) was used in a home interview to describe the outcomes to date for a random sample of survivors from each cohort. The relative valuation of health states was determined through utility measurements (on a random sample of parents of Hamilton school children). (See 5 below.)

The utility assessments were combined with the mortality and morbidity data in order to express the outcome of care in quality-adjusted life years.

3. Allowance for differential timing and uncertainty

A discount rate of 5 per cent was applied to costs, earnings and effects (life years and quality-adjusted life years) occurring in the future in order to convert these to present values.

Sensitivity analyses were performed on all results to determine the robustness of the findings to major changes in key factors. Specifically, the following four factors were each varied independently over the given range to determine the impact on the findings:

—discount rate from 0 to 10 per cent;
—life expectancy within the extremes of forecasts;
—loss to follow-up assuming all lost have major damage or all lost are normal;
—and utility values over their range of uncertainty.

4. Results and conclusions

'Neonatal intensive care increased both survival rate and costs. Among newborns weighing between 1000 and 1499 g the cost was $59,500 per additional survivor (at hospital discharge), $2900 per life-year gained, and $3200 per quality-adjusted life-year gained; intensive care resulted in net economic gain (undiscounted) but a net economic loss when future costs, effects and earnings were discounted at 5% per annum. Among those weighing between 500 and 999 g, the corresponding costs were $102,500 per additional survivor, $9,300 per life-year gained, and $22,400 per quality-adjusted life-year gained; intensive care resulted in a net economic loss. By every economic measure, neonatal intensive care of 1000–1499 g infants was superior to that of 500–999 g infants.'

The authors went on to point out that a 'judgement concerning the relative economic value of neonatal intensive care of very low birth-weight infants requires the economic evaluation of other health programmes by similar methods'.

5. General comments

This study embodies a number of methodological developments; notably in the area of estimating multi-attribute utility functions and in apportioning hospital overheads by the simultaneous equation method. (The reader is referred to the original study and supporting working papers for more explanation.)

47 Bredin, H. C. and Prout Jr., G. R. (1977). One-stage radical cystectomy for bladder carcinoma: operative mortality, cost–benefit analysis. *Journal of Urology* **117**, 447–51.

1. Study design

1.1. *Study question*

What are the costs of different treatments for bladder carcinoma?(a) (See 1.3 below.)

1.2. *Alternatives appraised*

Staged approaches *versus* a one-stage approach.

1.3. *Comments*

The authors analysed unstaged (one-stage) cystectomies performed over a five year period at one hospital to determine rates of morbidity and operative mortality. Since patient outcomes were similar to those reported in the literature for staged procedures, the authors went on to consider costs.

2. Assessment of costs and benefits

2.1. *Enumeration*

The total costs of hospitalization incurred by 123 patients, and hence the average cost of a cystectomy, were calculated. The cost of hospitalization included the cost of room and board, the actual operations, and the use of laboratory, pharmacy, and radiology facilities. The authors then estimated the costs of four different therapeutic methods, including the cost of hospitalization, additional radiology treatment, and lost production time. Changes in the cost of cystectomies over time were also analysed, and compared to the Consumer Price Index and the Medical Care Price Inex.

2.2. *Measurement*

Hospital accounts were available for 123 of the 127 patients, from which the costs were extracted. The authors did not give the source of the clinical information; presumably this was from patient records.

2.3. *Explicit valuation*

Market values were presumably used, although the authors did not give details of how they calculated their itemized costs, total costs, or estimated costs of alternative therapies.

160

3. Allowance for differential timing and uncertainty

Discounting was not performed. It would be relevant if the different staging of therapy meant that resources were committed at significantly different points in time in the alternative regimens being compared. However, a difference of one or two years in the timing of resource outlays would not cause discounting to have a large empirical impact.

No sensitivity analysis was performed.

4. Results and conclusions

The average hospital cost of a one-stage total cystectomy was US$8169; room, board, and operating costs comprising a large part. Using this hospital cost figure, the authors estimated the costs of the different methods of staging the treatment, including the cost of additional radiology time, and lost production time.

Unstaged cystectomy including 12 weeks lost production but no additional radiology, would cost US$10 600; unstaged cystectomy with additional radiology, diagnostic evaluation, and 26 weeks lost production would cost US$16 500; staged cystectomy with additional radiology, diagnostic evaluation and 28 weeks lost production would cost US$19 900; and short-stay ileal conduct, with intensive radiology and six weeks lost production would only cost US$7100.

5. General comments

The assessment of the therapeutic benefits were quite thorough, but there was little discussion of how the costs for the particular approaches were arrived at. Additional information on the increase in costs of the operation, and the sources of reimbursement were included, but these only serve to confuse the main issue. It was quite difficult to follow the descriptions the authors gave of how costs were derived; they appeared to be trying to do too much in one article, especially in the absence of a careful discussion of the costing methodology.

48 Brooks, R. G., Brown, M. G., Marsh, J. M., and Woodbury, J. F. L. (1981). Costs of managing patients at a Canadian rheumatic disease unit. *Journal of Rheumatology* **8**(6), 937–48.

1. Study design

1.1. *Study question*

What are the costs of managing patients (as inpatients and as out-patients) in a rheumatic disease unit (RDU)?(a)

1.2. *Alternatives appraised*

No alternatives were examined. The study was primarily concerned with costing methodology. (See 5 below.)

1.3. *Comments*

2. Assessment of costs and benefits

2.1. *Enumeration*

The study considered hospital costs only.

2.2. *Measurement*

Data were based on the treatments given to 194 inpatients and 433 outpatients. The central question was how to identify costs directly attributable to RDU patients and how to apportion joint costs. Since nursing costs were a large proportion of inpatient costs, a workload study was undertaken. (See the companion volume, Section 7.1.)

2.3. *Explicit valuation*

Market values were used in the main, and a value was imputed for hospital space by reference to market rates for office space.

3. Allowance for differential timing and uncertainty

Not considered. Given the necessity to make assumptions when apportioning costs, some sensitivity analysis of the cost estimates to changes in the assumptions would be advantageous.

162

4. Results and conclusions

The main result was that the cost of treating the RDU patients was lower than hospital inpatient *per diem* rates. Therefore it would be unwise to rely on routinely available hospital data to assess the total costs of any diagnostic sub-set of patients.

5. General comments

This study gives a good description of the problems of identifying costs within the hospital to particular patients. Its main weakness is that the costing exercise is not related to a particular question. There are a number of possible questions such as: 'what are the comparative costs of alternative regimens?' or 'what are the costs of expansion of the service?' Each question requires a slightly different approach to costing.

49 Churchill, D. N., Lemon, B. C., and Torrance, G. W. (1984), Cost-effectiveness analysis comparing continuous ambulatory peritoneal dialysis to hospital haemodialysis. *Medical Decision Making* **4**(4), 489–500

1. Study design

1.1. *Study question*

What is the most efficient form of renal replacement therapy for patients unsuitable for home haemodialysis or transplantation?(c)

1.2. *Alternatives appraised*

Continuous ambulatory peritoneal dialysis (CAPD) *versus* hospital haemodialysis.

1.3. *Comments*

The study question was addressed from the viewpoint of a provincial Department of Health in Canada.

2. Assessment of costs and benefits

2.1. *Enumeration*

The costs considered were restricted to direct costs incurred by the provincial Department of Health. These comprised all hospital and physician costs for the patients kept alive by the hospital haemodialysis or CAPD programmes, i.e. the costs of dialysis plus the costs of any other health services used by the patients. The only outcome considered was life years gained.

2.2. *Measurement*

The annual total for physicians fees for each patient was obtained from the Newfoundland Medical Care Commission. Hospital costs were allocated out to the appropriate cost centres. Annual programme costs for CAPD and hospital haemodialysis were divided by the life years gained by patients treated during the year to obtain the cost per life year gained. The patients were not matched or randomized in any way. Life years were not adjusted for quality differences but preliminary evidence was quoted to support the view that the outcomes were equivalent. (Churchill, D. N, Morgan, J., and Torrance, G. W. (1984). Quality of life in end-stage renal disease. *Peritoneal Dialysis Bulletin* **4**, 20–3.)

2.3. *Explicit valuation*

The opportunity cost of buildings and equipment was valued by calculating the return on the equivalent sum invested in the Newfoundland bond issue. All other resources were valued by their actual cost to the Department of Health.

3. Allowance for differential timing and uncertainty

All costs and outcomes occurred in the same year, with the exception of building and equipment costs. An annual equivalent for these was calculated by straight line depreciation at 7.5 per cent per annum for equipment and 2.5 per cent per annum for buildings.

Some sensitivity analysis was carried out on the level of nursing care required by the two groups when in hospital.

4. Results and conclusions

The average cost per life year gained for hospital haemodialysis and CAPD respectively were Can$48 700 and Can$33 400. If it is assumed that CAPD patients require more nursing care when inpatients and hospital haemodialysis patients less, the costs change to Can$47 000 and Can$41 000. If no hospitalization had been required in either group, the costs would have been Can$40 500 and Can$19 800. Despite the fact that the CAPD patients were older and sicker on average, the programme was more cost-effective than hospital haemodialysis.

5. General comments

The decision to include all health service costs and not just those attributable to dialysis or complications of dialysis was an unusual one and could have caused problems. Whilst the costs of treating coexisting diseases would clearly be relevant to a cost–benefit analysis, it should only affect the relative cost-effectiveness of different treatments if certain patients were more suitable for one regimen rather than another. In this case, the authors implied that all patients could be offered either therapy. However, the patients were not randomized and, in fact, those on CAPD were older and sicker. Since, despite this, the CAPD programme was more cost-effective the results are merely reinforced. In principle, the health service costs arising from non-dialysis related illness, which would have arisen irrespective of therapy chosen, could have confounded the results.

50 Culyer, A. J. and Maynard, A. K. (1981). Cost-effectiveness of duodenal ulcer treatment. *Social Science and Medicine* **15C**, 3–11.

1. Study design

1.1. *Study question*

What are the comparative costs of treating duodenal ulcer disease by a drug regimen and surgery?(c)

1.2. *Alternatives appraised*

A hospital-based surgical treatment (vagotomy) *versus* a community care drug treatment (cimetidine).

1.3. *Comments*

The alternatives considered in the study comprised two of many treatment alternatives available and were selected as they were the most frequently adopted and costly alternative treatments. The epidemiological evidence available at the time of the study was inadequate to permit comparison of drug and surgical intervention in terms of effectiveness. Indeed, the possibility was acknowledged by the authors that in some instances cimetidine may act as a complement to, rather than substitute for surgery, by delaying rather than preventing the need for surgical intervention.

2. Assessment of costs and benefits

2.1. *Enumeration*

In costing the surgical alternative, the authors included hospital costs, an estimate of loss of patient earnings attributable to surgery relative to cimetidine treatment and a value to take account of the risk of death associated with vagotomy. The authors conceded that no account was taken of pain, costs falling on patients' families, nor of costs falling on primary care or local authority services. The estimated case costs presented were acknowledged as being *minimal* cost estimates. In costing the cimetidine alternative, the only costs considered were drug costs. It was assumed for practical purposes that there would be no adverse side-effects and that the costs of supervision in the community would be minimal.

2.2. *Measurement*

The study was not linked to a randomized controlled trial and gathered data from various sources. Three alternative methods were used to estimate the hospital inpatient costs of surgery; multiple regression analysis using national data (see the companion volume, p. 70), application of national average length of stay data for vagotomy patients to the national average daily costs for acute hospitals, and direct observation of treatment provided for vagotomy patients in a major treatment centre — Newcastle Royal Victoria Infirmary.

Measurements of the differential patient time costs resulting from the alternative managements were proxied by loss of earnings estimates. Loss of earnings attributable to surgery relative to cimetidine were based on US data which reported differential times for return to work.

The estimate of case fatality consequent upon surgery was drawn from the literature.

In the absence of results from long-term cimetidine trials, the authors assumed that life-long drug maintenance would be required. For costing purposes two alternative dosages were assumed.

2.3. *Explicit valuation*

Where available, market prices were used and expressed at 1978 price levels.

In estimating the loss of earnings attributable to surgery relative to cimetidine treatment, a value for housewives' time was imputed using average wage rates (see the companion volume, p. 39).

Three different methods of assigning a value to human life were used in the calculation of case fatality associated with surgery; these were the social decisions approach, the human capital approach, and the value of risk avoidance approach (see the companion volume, pp. 40–3).

3. Allowance for differential timing and uncertainty

In calculating the costs of possible death resulting from surgery by the human capital approach, a 7 per cent discount rate over a 25 year period was used. This was the UK Treasury test discount rate at the time of the study.

In estimating the lifetime drug costs of maintaining a patient on cimetidine, the costs of 20, 25, 30, and 35 years of drug therapy were shown; in each instance, the present value of the required drug expenditure was calculated by discounting at 5, 7, and 10 per cent.

As the epidemiological information available for the study were rather weak and study data were collected retrospectively from various sources, the authors performed sensitivity analyses of the effects of major assumptions upon the results generated.

Culyer and Maynard (1981)

The authors attempted to load their calculations against the new procedure—cimetidine. They argued that by this means if the *highest* estimate of drug costs emerged as *cheaper* than surgery, then there should be some confidence that this was true.

4. Results and conclusions

With a 7 per cent discount rate when the highest and lowest cost estimates of vagotomy and cimetidine were compared there was a slight overlap, the highest estimate of the drug cost (£1240) exceeding (by £60) the lowest estimated cost of vagotomy ($1180).

The choice of discount rate was demonstrated to affect the relative costliness of the two procedures; at higher rates the drug regimen was shown to be cheaper than surgery, at lower rates there was some overlap.

At all discount rates there was a significant difference between the minimum and maximum vagotomy cost generated. This was attributable to the method used to value life in the estimation of the cost of case fatality.

From the perspective of the community as a whole, surgery was demonstrated to be the more expensive alternative. However, from the more limited perspective of the NHS, surgery would be preferred on cost grounds.

When the effects of the alternatives on patient time costs were excluded and no allowance made for possible case fatalities under the surgical alternative, the *highest* surgical cost was shown to be only approximately half that of cimetidine.

5. General comments

This study included a useful discussion of the alternative approaches to the valuation of human life. The authors also demonstrated the importance of the perspective adopted (patient, NHS, or society) upon the conclusions to be drawn from an economic evaluation. The study provided a very good example of the application of sensitivity analysis as an aid to overcoming problems of data weakness.

In conducting the study in the face of uncertainty of the effectiveness and costs of the two alternatives considered, the authors' intention was to weight calculations *against* cimetidine. However, in adopting this approach, the authors failed to attribute any adverse side-effects to long-term cimetidine.

51 Engleman, S. R., Hilland, M. A., Howie, P. W., McIlwaine, G. M., and McNay, M. B. (1979). An analysis of the economic implications of elective induction of labour at term. *Community Medicine* **1**, 191–8.

1. Study design

1.1. *Study question*

Does a policy of elective induction lead to a greater use of resources?(c)

1.2. *Alternatives appraised*

Elective induction at term *versus* conservative management.

1.3. *Comments*

The study was linked to a prospective randomized clinical trial involving 228 women. Significant differences between the two groups, in terms of resources and outcome, were the focus of attention.

2. Assessment of costs and benefits

2.1. *Enumeration*

Significant differences in the use of health service resources were considered. Other costs were not discussed, but it is unlikely that there would be differences between the two groups.

2.2. *Measurement*

Case records were the source of data on resource use and outcomes. The problem of joint costs was discussed, but the concentration on *differences* in resource use meant that it was not necessary to carry out any estimation of this type. (That is, in a cost-effectiveness analysis it is sometimes appropriate to eliminate costs that are *common* to all the alternatives being examined.)

2.3. *Explicit valuation*

Market values were used to estimate costs.

3. Allowance for differential timing and uncertainty

Differential timing was not relevant in the context of this study, since all the resource outlays under both alternatives occurred within a short space of time.

Engelman *et al.* (1979)

4. Results and conclusions

'The main conclusion was that apart from the resources involved in the induction process itself, induction of labour did not lead to a significantly increased use of hospital services. On the contrary, patients managed by a more "conservative" policy in the control group placed greater demands on the hospital service in the antenatal period in that they had significantly more attendances at the antenatal clinic.'

The net additional resource use by the trial group amounted to three hours of midwives time per week and approximately £1600 per year for supplies and equipment, both of which are negligible in the context of total hospital cost. No significant differences in outcome between the two groups were identified in the study.

5. General comments

52 Forster, D. P. and Frost, C. E. B. (1982). Cost-effectiveness study of outpatient physiotherapy after medical meniscectomy. *British Medical Journal* **284**, 485–7.

1. Study design

1.1. *Study question*

Should outpatient physiotherapy be provided for male patients, aged 16–45, after medial meniscectomy?(d)

1.2. *Alternatives appraised*

Outpatient physiotherapy *versus* no physiotherapy.

1.3. *Comments*

Both groups of patients received the same inpatient treatment and were instructed in exercise methods to use at home. Possible non-attendance by test group patients was accepted as reflecting the real state of the world, but did not prove to be a major factor in practice.

2. Assessment of costs and benefits

2.1. *Enumeration*

Outcomes were assessed by clinical aspects of knee function, the time taken to return to work and the total days lost from work. Time used to attend outpatient physiotherapy was included in days lost from work for the test group. The only other cost reported was for the provision of physiotherapy. Additional patient costs were considered but they were found to be small when compared with lost earnings and were not reported.

2.2. *Measurement*

The study was linked to a randomized controlled clinical trial. Measurement of knee function was made by two observers. Information on time lost from work and lost earnings was collected by means of a questionnaire. Details of the measurement of physiotherapy costs were not given but they were reported to be average costs.

2.3. *Explicit valuation*

The difference in work time lost was valued by the actual difference in take-home pay. Although it was not clear, it must be assumed that sickness benefits, etc. were netted out, in which case the difference in take-home pay cannot be equated with the difference in the value of lost output. (See the companion volume Section 2.1.1, on transfer payments.) Physiotherapy costs are assumed to have been valued at NHS wage and price levels, although this was not stated in the paper.

3. Allowance for differential timing and uncertainty

Differential timing was not relevant as all resource outlays under both regimens occurred within a short space of time.

4. Results and conclusions

There was no significant difference between the groups in any of the clinical aspects of knee function. The time taken to return to work and the total days lost were not significantly different for the two groups, nor was the average loss of take-home pay, which was approximately £162. The average cost of a course of physiotherapy was £23. The authors concluded that there was no significant advantage from routine out-patient physiotherapy for this group of patients. They also argued that the results should not be extrapolated to other age groups or conditions.

5. General comments

53 Foster, G. E., Bolwell, J., Balfour, T. W., Hardcastle, J. D., and Bourke, J. B. (1981). Clinical and economic consequences of wound sepsis after appendicectomy and their modification by metronidazole or providone iodine. *The Lancet* **i**, 769–71.

1. Study design

1.1. *Study question*

How effective is intrarectal metranidazole compared to interparietal providone iodine in the prevention and modification of wound sepsis following emergency appendicectomy? Is any higher effectiveness reflected in lower resource use and lower losses of income by patients?(c)

1.2. *Alternatives appraised*

The use of metranidazole *versus* providone iodine *versus* doing nothing.

1.3. *Comments*

Patients who received additional antibiotic therapy were withdrawn from the study.

2. Assessment of costs and benefits

2.1. *Enumeration*

The following clinical and economic consequences of the alternative procedures were considered: the incidence of clinical complications, associated length of stay, the number of post-discharge district nurse visits, loss of patient income during hospital stay and convalescence, and time taken to return to full activity.

2.2. *Measurement*

This study was part of a double-blind randomized controlled trial to which 496 patients over the age of 12 years were admitted. Randomization was so designed that the final ratio of treatment groups—metranidazole, providone iodine and control (no therapy) was 2:1:1.

Patient information relating to the post-discharge period was obtained from patients during follow-up out-patient appointments.

2.3. *Explicit valuation*

The derivation of the costs presented is uncertain. Presumably market values were used.

3. Allowance for differential timing and uncertainty

Not considered. Differential timing of costs and benefits was not particularly important in the context of this study.

4. Results and conclusions

Wound infection was significantly less common (p < 0.01) in patients receiving metranidazole (12.3 per cent) than in those receiving providone iodine (24 per cent) and in the untreated control group (23.5 per cent).

When the appendix was gangrenous or perforated, the wound infection rate in the metranidazole group was high (44 per cent), but less than that in the providone iodine group (62.5 per cent) and the untreated group (70.3 per cent); the difference between the metranidazole and control groups being significant (p < 0.05).

Patients receiving metranidazole left hospital earlier than those in the other two groups; the mean stay of the metranidazole, providone iodine and control groups being 5.1, 6.8, and 6.9 days respectively. Since wound sepsis was less common in the metranidazole patients, fewer district nurse visits were made to them than to patients in the other groups. As the wound infection rate was similar in the providone iodine and control groups, the authors suggested that the use of providone iodine for prophylaxis in emergency appendicectomy should be reappraised. Extrapolating their findings on loss of earnings and social security payments associated with the alternative regimens, the authors estimated that patients in England and Wales would have avoided losing about £1 million if metranidazole prophylaxis been universally used. Similarly, savings in social security payments might have been as much as £25 million a year.

5. General comments

The basis of the calculation of patients' financial losses during treatment and convalescence is uncertain. Whilst from the information presented on study design it appears that the calculation of NHS costs of the alternatives would have been feasible, such information is not presented. The authors do note, however, that the shortening of hospital stay associated with the adoption of metranidazole regimen may have the paradoxical effect of increasing costs because more patients are treated. Savings would only be expected if bed closures result. (See the companion volume, Section 7.1.)

Finally, it is not normal to include transfer payments (such as social security payments) in an economic appraisal. (See the companion volume, Chapter 2.)

54 Kay, H. E. M., Powels, R. L., Lawler, S. D., and Clink, H. M. (1980). Cost of bone-marrow transplants in acute myeloid leukaemia. *Lancet* **i**, 1067–9.

1. Study design

1.1. *Study question*

What is the cost of a bone marrow transplant and how does it compare with conventional therapy?(a)

1.2. *Alternatives appraised*

Bone marrow transplant for acute myeloid leukemia *versus* conventional therapy. (See 1.3 below.)

1.3. *Comments*

While the costs of the bone marrow transplant were examined in some detail the costs of the alternatives were examined only in approximate terms. The study does not constitute a detailed cost comparison.

2. Assessment of costs and benefits

2.1. *Enumeration*

The main costs considered were those resources directly used in the treatment and follow-up of patients (such as medical and nursing time, drugs, fluids, and radiotherapy). In addition, a daily charge for general hospital services was included. The authors also recognized that there would be capital costs in setting up a transplant unit but these were not included.

2.2. *Measurement*

The cost estimates were based on the resources used by the first 22 patients to be transplanted. Length of stay was recorded and items of extra cost (for the patient series concerned) identified and then added to the standard daily hospital cost (as reported in the hospital cost accounts). The main extra costs identified were in medical and nursing time, medical supplies, pathology, and radiotherapy.

2.3. *Explicit valuation*

Market values were used to estimate costs.

175

Kay *et al.* (1980)

3. Allowance for differential timing and uncertainty

Not considered. Differential timing is not particularly important in the case of the question being examined.

4. Results and conclusions

The basic cost of a bone marrow transplant was estimated to be £5564. This compared favourably with estimates for the United States but might be higher for transplants in less favourable circumstances (e.g. after relapse).

If one added, to the basic cost, the cost of treatment to bring the patient into remission, the cost of follow-up outpatient visits, and the costs of treatment for those patients who relapsed, the total figure was £11 564. This compared with £7500 for conventional treatment which may have a worse outcome.

5. General comments

This study, in costing bone marrow transplants, contributes to the debate over clinical alternatives in cancer therapy: who to treat, how to treat? However, it cannot alone answer these questions and a more detailed analysis of costs, risks and benefits, perhaps couched in a decision analysis framework, would be desirable.

Ludbrook, A. (1981). A cost-effectiveness analysis of the treatment of chronic renal failure. *Applied Economics* **13**, 337–50.

1. Study design

1.1. *Study question*

What is the relative cost-effectiveness of alternative methods of treating chronic renal failure?(c). (See 1.3 below.)

1.2. *Alternatives appraised*

Hospital dialysis *versus* home dialysis (with periodic hospital dialysis) *versus* transplantation (including home and hospital dialysis when required).

1.3. *Comments*

This paper also presents a methodological development in the assessment of the cost-effectiveness of chronic conditions. This uses a Markov model of transitions from dialysis to transplanted states and *vice versa*. The advantage of this approach over earlier cohort analysis is that the analysis can be easily updated as new outcome date become available.

2. Assessment of costs and benefits

2.1. *Enumeration*

The benefits considered were years of life gained. In addition there was some discussion of the *quality* of life gained, although the estimates were not adjusted to reflect differential quality of life on dialysis and transplant. The costs considered were the direct treatment costs under each regimen.

2.2. *Measurement*

Effectiveness data were drawn from the North East Thames Regional Health Authority (UK) and the European Dialysis and Transplant Association. These were used to calculate transition probabilities for each possible movement between states. The expected treatment profile for each package was then derived from the matrix of transition probabilities by calculating the expected length of stay in each state for a patient starting in State 1.

The expected treatment profiles were costed as if they were a sequential description of the treatment experience of an average patient undergoing each treatment process. No further details of costing were given.

Ludbrook (1981)

2.3. *Explicit valuation*

NHS expenditures were probably used to estimate costs. No further details were given.

3. Allowance for differential timing and uncertainty

Costs and effects were discounted at 7, 10, and 15 per cent. Two figures were used for each stage in the treatment process, representing high and low estimates of the actual cost.

In addition, two alternative assumptions were used to estimate age specific cost effectiveness results; (*i*) that transition probabilities for any age were proportional to those for all ages, and (*ii*) that the probabilities patients in the 15 to 34 age group have a transplant or go to home dialysis were increased by 10 per cent at all stages and that the same probabilities were decreased by 10 per cent for the 45 to 54 age group and by 20 per cent for the 55 to 64 age group.

4. Results and conclusions

The treatment package including transplantation ranked as the most cost effective in all age groups and at both discount rates for the first assumption, of proportional transition probabilities. The average treatment cost by this method varied from £3000 to £5000 per life year saved, depending on age, discount rate and cost assumptions. For the revised age assumptions, treatment including transplantation was still the cheapest, except for the highest discount rate and the older treatment groups. If it were assumed that the quality of life was higher with transplant, any ambiguity in the cost-effectiveness results was removed.

5. General comments

The paper also contains a discussion of policy implications of this kind of work and a comparison with estimates obtained by earlier analysts.

6 Mattson, W., Gynning, I., Carlsson, B., and Mauritzon, S.-E. (1979). Cancer chemotherapy in advanced malignant disease: a cost–benefit analysis. *Acta Radiologica Oncology* **18**, 509–20.

1. Study design

1.1. *Study question*

What are the costs and benefits of chemotherapy in advanced malignant disease?(c). (See 1.3 below.)

1.2. *Alternatives appraised*

In general: adjuvant chemotherapy *versus* no chemotherapy for all types of malignant disease. *Specifically*: two types of combination chemotherapy for metastatic breast carcinoma.

1.3. *Comments*

The study is mainly concerned with adding the chemotherapy modality to existing therapies. Therefore it is essentially a cost-effectiveness study.

2. Assessment of costs and benefits

2.1. *Enumeration*

The main costs considered were the costs of drugs and those associated with hospital inpatient stay. Benefits were expressed in terms of survival times and, to a lesser extent, ability to return to work and to lead a normal life.

2.2. *Measurement*

The general question of the impact of chemotherapy was investigated by way of a 'before and after' study (chemotherapy was introduced progressively from 1973). Changes in the number of patients treated, the mean number of days of inpatient stay and survival were noted.

The specific question was investigated through a randomized trial of the two combination therapies. Drug use, length of stay, and survival were noted.

2.3. *Explicit valuation*

Market prices were used to estimate the costs of drugs used in chemotherapy. The approach to valuing the other resources was to argue that, owing to chemotherapy, the hospital had been able to treat more patients

179

Mattson *et al.* (1979)

with its available staffed beds. This was because chemotherapy reduced inpatient stays. No attempt was made to value the increased survival owing to chemotherapy.

3. Allowance for differential timing and uncertainty

Not considered. Strictly speaking it would have been relevant to discount costs and benefits as adjuvant chemotherapy may delay the progression of the disease, thereby postponing the costs of terminal care. However, the *empirical* impact of discounting is likely to be small.

4. Results and conclusions

The advent of chemotherapy was the main change in the treatment practices of the department over the years 1973 to 1977. The extra cost in drugs was US$190 000. This compared with an increased throughput of patients which was accomplished with no extra beds or staff, and increased survival and quality of life.

The relative costs of drugs, length of hospital stay and survival of patients on the two combination therapies were presented, although the paper did not clearly state that one was preferable to the other.

5. General comments

Oberle, M. W., Merson, M. H., Shafiqul Islam, M., Mizanur Rahman, A. S. M., Huber, D. H., and Curlin, G. (1980). Diarrhoeal disease in Bangladesh: epidemiology, mortality averted and costs at a rural treatment centre. *International Journal of Epidemiology* **9**(4), 341–8.

1. Study design

1.1. *Study question*

What are the costs and outcomes (in terms of deaths averted) for treating diarrhoeal disease in a rural treatment centre?(d)

1.2. *Alternatives appraised*

Rural treatment centre *versus* (implicitly) 'doing nothing'. (See 1.3 below.)

1.3. *Comments*

The question of an alternative programme was not really addressed. It was tacitly assumed that many patients would die without rehydration therapy and that there was no practical alternative method of delivering the therapy. (See 4 below.)

2. Assessment of costs and benefits

2.1. *Enumeration*

The costs considered were those for personnel and medicine, including intravenous and oral rehydration fluid, speedboat and automobile transport, and building rental and maintenance. Not included were annuitization of earlier equipment purchases and costs for the small outpatient department and for clinical studies. The outcome measure considered was 'deaths averted'. (See 5 below.)

2.2. *Measurement*

Effectiveness of the programme was estimated by a retrospective review of hospital records since 1975. It was argued that patients with a given status on admission would have died had therapy not been available. (High and low estimates of deaths averted were obtained.)

2.3. *Explicit valuation*

Market prices (operating expenditures) were used to calculate costs.

181

Oberle *et al.* (1980)

3. Allowance for differential timing and uncertainty

No discounting was employed, but given the duration of the therapy being evaluated, this does not represent a serious problem for the analysis.

No comprehensive sensitivity analysis was performed although, as indicated in 2.2 above, high and low estimates of deaths averted were made.

4. Results and conclusions

It was estimated that approximately 25–50 per cent of the hospitalized patients would have died had no rehydration therapy been available. The region's total mortality was reduced by about 7–15 per cent at a cost of US$0.14 per capita. (The average cost per death averted was US$38–81.)

The authors acknowledged that, with growing experience in oral rehydration therapy, less intravenous therapy was being used in the hospital. A community-based oral rehydration therapy programme had been implemented in the area and these data 'do not address the question of whether a community-based or a hospital-based programme is the most effective mechanism for the delivery of rehydration treatment'.

5. General commands

The authors noted that, although the costs per death averted were much lower than those for new therapeutic approaches in developed nations, the entire health budget of Bangladesh is estimated to be only US$1.00 per capita, just seven times this hospital's expenditure.

The main defect in the study is the lack of consideration of an alternative programme. From the discussion presented in the paper it seems that the most likely alternative would be a community-based oral rehydration programme rather than an immunization programme.

In common with many economic evaluations carried out in developing countries, this study measures 'deaths averted' rather than 'years of life gained'. This is important since the risks from other causes of death are quite high in such a setting.

8 O'Donnell T. F., Gembarowicz, R. M., Callow, A. D., Panker, S. G., Kelly, J. J., and Deterling, R. A. (1980). The economic impact of acute variceal bleeding: cost-effectiveness implications for medical and surgical therapy. *Surgery* **88**, 693–701.

1. Study design

1.1. *Study question*

What is the relative cost-effectiveness of medical and surgical procedures for treating acute variceal-bleeding?(c)

1.2. *Alternatives appraised*

Emergency medical treatment *versus* emergency surgical treatment *versus* elective medical treatment *versus* elective surgical treatment.

1.3. *Comments*

All patients had the same initial treatment on admission. The elective therapy choices applied only after acute bleeding had been controlled. Patients for whom bleeding could not be controlled underwent emergency treatment. Also, three elective medical patients were transferred to the elective surgical group after complications arose. The entire study covered 32 patients.

2. Assessment of costs and benefits

2.1. *Enumeration*

The only costs considered were the hospital costs for the initial episode of treatment. Subsequent health care costs were ignored, as were costs falling on the patient, family, or elsewhere in the economy.

The effectiveness comparison was based on survival time, the recurrence of bleeding and other complications, the numbers returning to work and the quality of life.

2.2. *Measurement*

The study was not linked to a randomized controlled trial, but reported the results of other controlled studies as well as the results for the study patients. Survival times for the study patients were extrapolated from the follow-up by assuming a constant annual mortality rate. Hospital resource use was itemized from the patients' bills, thereby reflecting average rather than marginal costs.

O'Donnell *et al.* (1980)

2.3. *Explicit valuation*

Hospital charges were used to value costs. The benefits from different rates of return to work and the averted costs of subsequent hospital treatment were not valued.

3. Allowance for differential timing and uncertainty

Discounting was not discussed, although it was relevant to this study.

The authors were aware of the limitations of the data from their own study and used the results of other studies to provide alternative cost-effectiveness estimates.

4. Results and conclusions

No patient in either of the emergency groups survived initial hospitalization. The cost per life year saved for the elective surgical group was US$30 371 and for the medical group was US$24 000. If the survival data from other studies were applied, these figures were reduced to US$15 206 and US$8226 respectively.

However, more of the surgical patients returned to work; eight out of nine compared with four out of 11 medical patients. Medical patients also had more re-admissions to hospital; a total of 13, compared with one for the surgical group. The cost-effectiveness results were recalculated to take account of time spent in hospital, as a proxy quality of life adjustment. This showed surgical treatment to be more cost-effective, costing US$2432 per month out of hospital, compared with US$3592 for the medical group. The authors concluded that the higher costs of initial surgical treatment may be offset by fewer re-admissions and improved quality of life.

5. General comments

The results of the study cannot be considered conclusive. The patients were not randomized or matched and, therefore, the severity of illness may have varied. All relevant costs were not included and there was no discounting. Discounting would have reduced the cost-effectiveness of the surgical treatment, compared to the medical alternative, as many of the benefits of surgery occurred in the future, due in part to the longer survival of surgical patients. However, the use of time spent out of hospital as an adjustment for quality of life was a novel concept.

Pentol, A., Allen, D., and Maguire P. (1981). The hidden cost of a mastectomy. *General Practitioner* 29.

1. Study design

1.1. *Study question*

What are the costs and effectiveness of the provision of counselling by psychiatrically trained nurses for mastectomy patients?(d)

1.2. *Alternatives appraised*

Routine care *versus* routine care plus help from a nurse counsellor during the pre- and post-operative period.

1.3. *Comments*

2. Assessment of costs and benefits

2.1. *Enumeration*

In comparing the costs of the alternatives, the authors considered costs incurred by the NHS (e.g. psychiatric inpatient treatment, district nurse time, time spent with patients' GPs, the costs of drugs prescribed, and the cost of providing the counselling service), the costs of local authority services (e.g. home helps, meals on wheels, and social workers), costs incurred by the patients themselves, and costs incurred by patients' family and friends.

The effectiveness measure used in evaluating the provision of nurse counselling was that of the incidence of psychiatric and social problems exhibited in patients who were offered counselling compared to that of those who were not.

2.2. *Measurement*

The effectiveness of the provision of nurse counselling was measured within the framework of a randomized controlled trial. A group of 152 women undergoing mastectomy were randomly assigned to a programme of routine care plus help from a nurse, or routine care alone. Independent assessors determined the incidence of psychiatric and social problems exhibited by women in the two groups during a 12–18 month follow-up period, using standard assessment methods. The economic implications of the two alternatives were measured retrospectively; the means by which these data were collected is not described in this article.

Pentol *et al.* (1980)

2.3. *Explicit valuation*

Presumably market prices were used where available. Three alternative methods were used to value the domestic work which patients would have carried out at home had they been well. One method was to assume that housewives' labour had no value; another method was to value housework at the average wage rate of working women in the study, and the third method was to assume that the value of housework lost was equal to the value of work done by women employed outside their homes—this final method of valuation was combined with the assumption that the length of time housewives were unable to work in the house would, on average, equal the time lost from work by those females who were employed.

3. Allowance for differential timing and uncertainty

Not considered. Differential timing was not particularly relevant in the context of this study.

4. Results and conclusions

At follow-up the authors reported that there was 'much less' psychiatric and social morbidity in the group receiving nurse counselling than in the control group, although the extent of this difference was not quantified.

The costs of counselling in the experimental group were partially offset by the costs of the higher re-admission rate of psychiatric treatment for the control group, but the average NHS cost per case in the experimental group slightly exceeded that of the control group; the NHS costs being £116 and £99 respectively.

Local Authority costs for both groups of patients were identical at £19 per patient. The costs incurred by patients varied depending upon the method by which housework was valued, but in all cases the average cost incurred by patients in the experimental group exceeded those incurred by patients in the control group.

Costs incurred by the relatives of patients allocated to the control were on average three times as great as those incurred by families of patients referred for nurse counselling.

The average total cost incurred by patients receiving counselling was lower than the average total cost incurred by patients in the control group, except where the assumption was made that the value of work in the home was equal to that of women employed outside the house. Depending upon the method used to value housework, the average total cost incurred by women in the experimental group was £809, £977 or

£1448; with the corresponding costs for women in the control group being £928, £1012 or £1404. Hence, from the perspective of the NHS, the provision of nurse counselling would lead to an increase in the costs associated with mastectomy, whereas from the perspective of society, assuming that housework had a zero value or a value no greater than that of the domestic wage rate paid to domestics, then the provision of nurse counselling would lower the costs of mastectomy.

5. General comments

This study is presented in greater detail in: Pentol, A. and Allen, D. (1981). *The costs of breast cancer*. University of Manchester Health Services Management Unit Working Paper No. 42. Also a more readily available report on the study is given in: Maguire, P. *et al.* (1982). Cost of counselling women who undergo mastectomy. *British Medical Journal* **284**, 1933–5.

It would have been helpful to have had information relating costs and outcome to compliance with the counselling service. Furthermore, a development of the study would have been to identify sub-groups of mastectomy patients whose situation and case history suggested that they would derive particular benefit from a counselling service; it might be that the service would result in a reduction in NHS costs if provided for such high risk patients.

This study indicates the high sensitivity of results to different methods of valuing household labour.

60 Ricardo-Campbell, R., Eisman, M. M., Wardell, W. M., and Crossley, R. (1980). Preliminary methodology for controlled cost–benefit study of drug impact: the effect of cimetidine on days of work lost in a short-term trial in duodenal ulcer. *Journal of Clinical Gastroenterology* **2**, 37–41.

1. Study design

1.1. *Study question*

Does the use of cimetidine in treating duodenal ulcer reduce time lost from work?(b)

1.2. *Alternatives appraised*

Cimetidine *versus* placebo, with both groups taking antacids as desired for symptom relief.

1.3. *Comments*

The study was a pilot project, intended to illustrate the use of economic appraisal in drug trials, rather than to provide a full evaluation of cimetidine. The study was short term and did not consider cimetidine *versus* surgery for longer-term management.

2. Assessment of costs and benefits

2.1. *Enumeration*

The costs for which data were desired but not collected included treatment costs and adverse reaction costs. A complete enumeration of costs was not given. The benefit considered was the reduction in time lost from work. Gains in non-marketed healthy time were not included (see comments in the companion volume, Section 4.1). Quality of life was mentioned briefly but was considered to require 'psychological or sociological evaluation'.

2.2. *Measurement*

Only the improvement in time lost from work was actually measured. Data were obtained by direct questioning of patients.

2.3. *Explicit valuation*

No monetary values were attached to the benefits. The individuals' actual wage rates would have been used, if available.

3. Allowance for differential timing and uncertainty

The question of differential timing did not apply in the context of this study.

4. Results and conclusions

The difference in time lost from work for the treated and control groups prior to the start of the trial was not significant. The group treated with cimetidine showed a greater improvement in time lost during each treatment period and the results were statistically significant at either 5 or 1 per cent test levels. However, the results were sensitive to assumptions made about missing data for eight patients.

5. General comments

The distribution of days lost from work, both before and after treatment, showed that most people either worked a full week or were absent for the whole week. It was noted that this could be due to institutional factors and the distribution of data on the patients' 'overall feeling' tended to confirm this. The study numbers were fairly small, with a total of 64 patients taking part. Of this group, 20 were followed for a six-week period, a further 26 were studied for four weeks, and the rest were studied for only two weeks.

61 Roberts, S. D., Maxwell, D. R., and Gross, T. L. (1980). Cost-effective care of end-stage renal disease: a billion dollar question. *Annals of Internal Medicine* **92**(1), 243–8.

1. Study design

1.1. *Study question*

What is the relative cost-effectiveness of alternative forms of treatment for end-stage renal failure?(c)

1.2. *Alternatives appraised*

Home haemodialysis *versus* centre haemodialysis *versus* live related donor transplant *versus* cadaver transplant.

1.3. *Comments*

A common problem with studies in this area is the difficulty of taking into account newly developing forms of treatment. Thus, this study does not include continuous ambulatory peritoneal dialysis (CAPD), which would now be considered a relevant alternative.

2. Assessment of costs and benefits

2.1. *Enumeration*

The costs considered were average health care costs only. Costs to the patient, their family, or elsewhere in the economy were not included, but installation and conversion costs were included for home dialysis. Life years gained were used for the comparison of effectiveness. No adjustment was made for differences in the quality of life under alternative treatments.

2.2. *Measurement*

The treatment patterns for a cohort of 10 000 patients were simulated from actual patient data. Most of these data were taken from the reports of the National Dialysis Registry and the Human Renal Transplant Registry. Average costs of treatment episodes were used to cost the treatment patterns.

2.3. *Explicit valuation*

Details of the costs used were given in the paper. They were based on national data for medical fees and other costs, and on the reimbursement levels set under Medicare.

3. Allowance for differential timing and uncertainty

The results were first presented without discounting. As part of the sensitivity analysis, the effect of discounting costs at 5 per cent and benefits (life years) at 7 per cent was shown and the authors report that the results were not substantially affected by discounting over a range of rates (0–20 per cent). The sensitivity analysis also examined the effect of varying the survival probability on the cost-effectiveness of home dialysis, standardizing the age mix of the patients in different treatment regimes and the effect of longer survival times for cadaver transplants. The effect of improvements in dialysis and transplant technology makes such sensitivity analysis particularly important, in compensating for the necessary use of historic survival rates in estimating models.

4. Results and conclusions

The undiscounted results gave live related donor transplant as the most cost-effective form of treatment at US$7709 per life year saved. The cost per life year gained for the other treatment types was US$13 274 for home dialysis, US$14 918 for cadaver transplantation and US$24 800 for centre dialysis. The ranking of cadaver transplant and home dialysis was reversed under the assumption of longer survival times for cadaver transplants.

5. General comments

Other studies in this area have tended to support transplantation as the most cost-effective treatment, regardless of donor type. The discrepancy may be due to any of three factors; the survival times, the estimation procedure, or differences in the relative prices of treatments. Certainly, the sensitivity analysis showed that using improved survival times for cadaver transplants made this option more cost-effective than home dialysis. However, relative price differences for treatments may be particularly important when comparing results from studies carried out in the UK and the United States.

Two further points should be noted. Although the authors use different discount rates for costs and benefits, they fail to give any rationale for this. Implicit in this procedure is the view that there will be a change in the relative values of costs and benefits equivalent to 2 per cent per annum. Secondly, when discussing the policy implications of their results, the authors do not pay sufficient attention to whether the results

191

of their average estimations can be applied to change in the mix of treatments at the relevant margin. (For example, because of difficulties in obtaining kidneys, the costs of expanding the cadaver transplant option may be high at the margin, as publicity campaigns may have to be mounted.)

2 Rundell, O. H., Jones, R. K., and Gregory, D. (1981). Practical benefit–cost analysis for alcoholism programmes. *Alcoholism: Clinical and Experimental Research* **5**(4), 497–509.

1. Study design

1.1. *Study question*

What are the costs and benefits of alcoholism treatment programmes?(c) and (d)

1.2. *Alternatives appraised*

A representative treatment programme *versus* an average of 26 programmes in Oklahoma *versus* (implicitly) no programme.

1.3. *Comments*

In previous papers, the authors had discussed the results of a cost–benefit analysis of the entire state-wide system of alcoholism treatment programmes in Oklahoma and the methodological issues. In this paper, they have selected a representative treatment programme (designated programme A) and have given a detailed account of the procedures used and the results obtained.

During the study period 346 clients were admitted to programme A. Data were collected at admission and at six month follow-up, by interview and questionnaire, and also from the state computerized management system.

2. Assessment of costs and benefits

2.1. *Enumeration*

The analytical perspective was that of the national economy. Changes over the six month period for the clients, and where appropriate in the economy, were monitored.

Benefits were measured by increases in productivity, improvement in health and reductions in automobile accidents, arrests, and criminal justice system costs. Costs of services were based on reimbursement rates, averaged out by client, in the absence of unit cost information.

2.2. *Measurement*

All benefits were expressed in money terms. Data collected at the two measurement points were converted to annual rates to give pre- and post-treatment data, and the gains (or losses) by benefit category calculated by client.

Increased productivity was measured by the change in earnings, modified by inflation rates, and an income multiplier which took into account the expected effect on the economy. Health benefits were measured by the changes in resource use resulting from a change in the number of hospital visits and use of other services. Reduced automobile accident losses took into account the reduction in premiums to insurance companies, changes in risk categories, and the proportion of clients who drive. Reduced criminal justice system costs took account of the reduction in the number of arrests and the associated costs.

The costs of services considered included those for intermediate (residential) care, medical and non-medical detoxification, out-reach services, client evaluation, individual, group and family counselling, and all follow-ups. An average cost per client was calculated.

2.3. *Explicit valuation*

Costs were pro-rated, based partly on reimbursement fees established for each service by the State Department of Mental Health, and partly on the budgets set for the separate agencies and for the State Department Division of Alcoholism.

Income levels were ascertained by questionnaire; hospital costs were taken as the daily average for Oklahoma; and insurance rates were current market rates. The authors did not give the method of calculating the costs of arrests.

3. Allowance for differential timing and uncertainty

A discount rate of 10 per cent was applied to costs occurring in the future. An inflation rate of 4 per cent over six months was used in the calculation of productivity benefits.

4. Results and conclusions

The total benefit per client over one year for programme A were £1428 and treatment costs US$321, yielding net benefits of US$1107, or a benefit–cost ratio of 4.4:1. The total benefits per client over the average of the 26 state programmes were US$1352 and treatment costs US$568, yielding net benefits of US$784, or a benefit–cost ratio of 2.4:1.

Six programmes were selected from the 26 for further analysis, yielding net benefits ranging from US$3316 to − US$3845 per client, or benefit–cost ratios between 7.6:1 to − 5.4:1.

The authors also offered suggestions for improvement in measurement and discussed the partial nature of the results, given the omission of intangible benefits.

5. Comments

This was a detailed study, particularly in the estimation of benefits. The cost estimation procedures were a little weak; given the care devoted to the methodology, it is surprising that the unit costs were not collected during the study.

63 Stange, P. V. and Sumner, A. T. (1978). Predicting treatment costs and life expectancy for end-stage renal disease. *The New England Journal of Medicine* **298**, 372–8.

1. Study design

1.1. *Study question*

What are the cumulative 10 year direct medical costs and life expectancy associated with different methods of treatment for end-stage renal disease?(c)

1.2. *Alternatives appraised*

Hospital dialysis *versus* treatment regimens incorporating various combinations of hospital dialysis, home dialysis, and transplantation.

1.3. *Comments*

The cumulative costs and life years associated with each method of treatment were estimated for successive annual cohorts of 1000 patients over a 10 year period. The predicted total 10 year experiences of cohorts changing from hospital to home dialysis, hospital dialysis to transplantation, and home dialysis to transplantation were then compared with the predicted experiences of the same aggregated cohorts in the absence of any change of treatment method.

The study excluded those patients who would not have stabilized by the end of the first 12 months of dialysis and hence who would not be medically eligible for all three treatment options.

2. Assessment of costs and benefits

2.1. *Enumeration*

Cost considerations were limited to medical costs and the cost of home adaption in the case of the home dialysis alternative. Patient costs and costs incurred by their family and other non-medical carers were excluded. The reliance of both the home dialysis option and transplantation upon hospital dialysis and other hospital support services was acknowledged.

The only benefit measure considered was that of additional years of life; the study does not concern itself with the issue of the different *quality* of life associated with the treatment alternatives.

196

2.2. *Measurement*

The study was not linked to a controlled clinical trial. The 10 year survival and cost data were obtained through linear extrapolation of recent trends reported in the literature. Survival estimates for patients allocated to a dialysis programme were based on those published in the 1976 report of the National Dialysis Registry. Those for patients undergoing renal transplant were based on estimates published in the 1976 Report of the Human Renal Transplant Registry of the American College of Surgeons/National Institutes of Health.

2.3. *Explicit valuation*

The derivation of the medical costs employed in the study is uncertain. It is likely that market prices were used.

3. Allowance for differential timing and uncertainty

A discount rate of 7 per cent was applied both to future costs and to life years gained; the case for discounting the latter being that 'an additional year of life now is more valuable than one in the future, which, owing to uncertainty, may not be experienced as predicted'. Prior to discounting, costs were increased by 2 per cent to reflect anticipated real cost increases due to such factors as the use of more advanced and more expensive technology, and general improvements in the quality of care.

A high and low survival estimate was employed in the assessment of the life expectancy of transplant patients.

4. Results and conclusions

Over a 10 year period the number of life years for a single cohort of 1000 patients assigned to home and hospital dialysis was predicted to be approximately the same, but home dialysis was predicted to cost US$43 million less than hospital dialysis. Transplantation was predicted to give rise to between 6 and 20 per cent less life years and to cost between US$26 million and US$67 million less than dialysis, depending upon the life expectation assumptions made and the type of dialysis with which transplantation is compared.

Similarly, as regards the treatment transition options considered, it was predicted that if 1000 patients changed from hospital to home dialysis for each of 10 years, life expectancy of the cohort would not be reduced, but costs would be reduced by US$241 million. The same number of patients changing from hospital dialysis to cadaveric transplantation was predicted to lead to a reduction of between US$279 million and US$330 million in costs, but also a reduction of between 7 and 17 per cent in life expectancy. The transition from home dialysis to transplantation was predicted to reduce costs by between US$104 million and US$142 million and life expectancy by between 10 and 20 per cent.

5. General comments

This study represents an advance upon many of its predecessors in that, in the comparison of alternative treatments for patients with end-stage renal disease, it is acknowledged that both home dialysis and transplantation require hospital dialysis and other hospital-based support services.

It is to be emphasized that the results of this study are only applicable to those segments of the population able and willing to undergo alternative forms of treatment; for many patients with end-stage renal failure there may be no practicable alternative to hospital dialysis. No estimate was made of the likely magnitude of savings resulting from patients transferring from hospital to home dialysis or to transplantation. The authors argue that to determine the savings for any constant annual volume of patients transferring from hospital dialysis, one need only multiply the cumulative cohort totals presented by the appropriate fraction (i.e. if 50 per cent transferred in practice, then the savings would be 50 per cent of the sums presented). However, it is possible that the savings per patient transferred from hospital dialysis might be affected by the total number of patients transferred, depending on the local arrangements for treatment provision.

Weinstein, M. C. (1980). Estrogen use in postmenopausal women—costs, risks and benefits. *New England Journal of Medicine* **303**, 308–16.

1. Study design

1.1. *Study question*

Is the exogenous oestrogen treatment of postmenopausal women worthwhile?(d)

1.2. *Alternatives appraised*

Oestrogen treatment *versus* doing nothing.

1.3. *Comments*

Three sub-populations were considered; symptomatic women treated from age 50–60, women with symptomatic osteoporosis treated from age 55–70, and asymptomatic women treated from age 50–65.

2. Assessment of costs and benefits

2.1. *Enumeration*

Costs and benefits were divided into three types. First, there was the net change in medical resources costs (i.e. the costs of treatment, side-effects of treatment, and additional cancer cases less the avoided costs of fractures). Secondly, there was the net change in life expectancy; and finally there was the net effect on the quality of life. Patients' costs and the effect on the productive output of treated women were not considered.

2.2. *Measurement*

Most of the data were taken from the medical and epidemiological literature. The quality of life adjustments were hypothetical subjective valuations. (See the methodological introduction to this volume.)

2.3. *Explicit valuation*

Treatment costs were based on prevailing fees and charges. Life expectancy changes and the quality of life were not explicitly valued in money terms, but presented as 'quality-adjusted life years'.

Weinstein (1980)

3. Allowance for differential timing and uncertainty

Costs were discounted at 5 per cent per annum. Benefits (gains in life expectancy) were presented with and without discounting, reflecting the author's view that discounting life years is controversial. The importance of sensitivity analysis was stressed and sample results were given for variations in the risk of cancer, and for the assumptions of no reduction in the risk of fractures and that the introduction of an annual biopsy reduces cancer mortality by half.

4. Results and conclusions

Net treatment costs were lowest for women with symptomatic osteoporosis (US$471 without annual biopsy, US$1300 with annual biopsy) and highest for asymptomatic women (US$695 and US$1753 respectively).

Changes in life expectancy were small. For the osteoporosis patients, the gain was 11 days (three days if discounted). For the other two groups the effect was smaller and negative. The value of the treatment depended on the qualitative benefits from symptom relief. A hypothetical value of 0.01 quality-adjusted life year gained per year of treatment was used to calculate cost-effectiveness results. Under central assumptions, the treatment of osteoporosis patients cost US$5500 per quality-adjusted life year gained. The corresponding figures for symptomatic and asymptomatic women were US$7400 and US$22 500 respectively. The treatment of women with a prior hysterectomy was more cost-effective as the risk of cancer is lower for these patients.

5. General comments

SECTION 5: Alternative locations of care

Introduction

The studies reviewed in this section concern the care of patients in different settings. Many of them examine the choice between institutional and community care, for example, should patients be cared for in hospital or in their own homes? There has been a strong interest in community care in most countries, partly because such care may be preferred by the patient and partly because it may be less costly. The studies therefore include some, in the acute field, dealing with the substitution of post-operative care in hospital by care at home (Evans and Robinson 1980; Prescott *et al.* 1978); some dealing with shifts towards community-orientated care for some mental illness patients (Fenton *et al.* 1982; Weisbrod *et al.* 1980), and some, in the care of elderly patients, comparing the costs of care in different settings (Wright *et al.* 1981) and evaluating schemes specifically designed to avert institutionalization of elderly or otherwise dependent persons (Gibbins *et al.* 1982; MacFarlane *et al.* 1979; Mowat and Morgan 1982; Weissert *et al.* 1980).

Other studies reviewed are concerned with not only the location of care but also the nature of the care given and the types of professionals involved. For example, Bloom and Kissick (1980) compare home and hospital care for terminally ill patients; Linn *et al.* (1979) compare costs and outcomes in hospitals with and without specialist burn units; Mangen *et al.* (1983) examine the relative cost-effectiveness of care given by psychiatric nurses in the community and psychiatrists in out-patient clinics; two studies compare GP and specialist maternity units (Gray and Steele 1981; Stillwell 1979); and Logan *et al.* (1981) examine the relative cost-effectiveness of hypertension treatment delivered at the worksite by nurses and treatment given in the community by physicians in private practice.

Finally, two studies (Chamberlain 1980, on family planning services and Kriedel 1980, on epilepsy clinics) examine the merits of specialist services and Doherty *et al.* (1980) seek to examine *why* dental care costs may differ among three practice settings (public fixed and mobile clinics, and private practice).

Particular methodological problems in this area

One of the main problems faced by analysts in this area is that of specifying the alternatives for appraisal. Often one is not dealing with a straightforward choice between (say) community care and institutional care, but a *mix* of such services, based on the level of dependency of the patient. Whereas the inclusion criteria for patients admitted to a study of therapeutic alternatives (Section 4) may be closely defined by reference to clinical indications, this may not be the case here. Often a community programme may be set up with the objective of accepting a broad range of patients; this begs the questions of 'what are they suffering from?' and, 'how bad is their condition?' and this makes the results of studies hard to interpret. Therefore it is important that studies in this area give explicit details of the principal diagnoses of patients enrolled in the community programme and undertake some degree of dependency assessment of those patients. Only then can clinicians and planners reading such studies form a clear view on the viability and relevance of such programmes in their own location.

Another difficulty in this area arises from the fact that in the health care systems of many countries the community care programmes are operated by agencies independent from those running institutional care. This raises difficulties in the implementation of change based on study results (discussed below) and in study design. A major implication is that it is important to identify costs and benefits *by viewpoint*. Also it may mean that comparative costs of care in different settings should not be taken at face value; they may reflect different payment or accounting conventions in the different agencies and not differences in real resource costs (the proper concern of economists). One useful ploy, therefore, would be to ensure that the physical resource inputs to care are listed for each of the care settings, as well as the global cost figures. In this way one might be able to distinguish between, on the one hand, the ways in which the various inputs are combined to provide care and, on the other, the returns to each factor. (The paper by Doherty *et al.* 1980, explores this issue in more detail and might provide some food for thought for those analysts particularly concerned with this question.)

Finally, this area highlights two methodological problems common to all economic appraisals, namely the valuation of production gains/losses and the valuation of unpriced inputs such as family helpers' time and volunteer time. The appropriate treatment of production gains/losses arises in this area because community-based treatments often enable patients to participate in the workforce to a greater degree than is possible in the institutional alternative. (Conversely they may also require that relatives have to forgo work opportunities to a greater extent.) In some cases the inclusion of production gains/losses can change the study result, as in the study by Weisbrod *et al.* (1980) of an experimental community-based programme for mental illness patients. The methodological problems of valuing production gains/losses have been discussed at length elsewhere (Drummond 1980; Williams 1981) and in the methodological introduction to this volume (Chapter 2).

Turning to the question of unpriced inputs, it is surprising that analysts do not pay more attention to the question of volunteer time and family time used in community care options. It might be found that institutional care, for example, consumes a significant amount of family time (in visiting hospitalized relatives) as well as the community-based alternative. Valuation of inputs of family time is complex and, to our knowledge, currently unresolved in the literature. Evans and Robinson (1980) argue for a revealed preference approach, that is, to offer the community-orientated service where this represents a good use of health care resources and to see whether people use the service. Alternatively, more effort could be put into interviewing patients and their relatives and exploring the level of compensation they would require (in improved support services or in financial help) for them to accept the community care option. Some of the studies reviewed here attempt such surveys, although not always in a systematic fashion.

Current state of the art

A number of potential improvements in study methodology in this area were suggested above. Here, two serious defects in the existing literature are discussed in more detail and gaps in the range of current applications of economic appraisal identified.

The two major defects are (*i*) a lack of controlled medical evaluation, and (*ii*) inadequate investigation of institutional (e.g. hospital) costs. These are important since most of the studies attempt to establish whether or not community care is more cost-effective than the institutional alternative. The paucity of randomized controlled trials in this area is partly explained by the greater organizational difficulties; compared to therapeutic trials in the hospital, they require the co-ordination of many more agencies. However, on the other hand they may raise fewer ethical concerns than (say) a trial of surgical *versus* medical management of angina. Nevertheless, the group of studies contains three examples of large randomized trials (Fenton *et al.* 1982; Weisbrod *et al.* 1980; Weissert *et al.* 1980), so these are clearly feasible. In particular the studies of Mowat and Morgan (1982), MacFarlane *et al.* (1979) and Gibbins *et al.* (1982) would have been strengthened had they incorporated a controlled medical evaluation.

In situations where such evaluations are not possible, more effort should be made to assess the dependency of patients in the different forms of care being compared, to standardize for the patients' condition. See, for example, the approach adopted by Wright *et al.* (1981) in their costing study of options for care of the elderly.

In the consideration of costs, most analysts investigate in detail the costs of the alternative to institutional care (usually in the community). This is because the community care option is the new or experimental programme of interest. Some analysts even discuss whether the new programme is operating in its most efficient manner, in terms of the *size* of operation, the *intensity of use* of facilities and whether *minor modifications* can be made to the procedures in order to make them more cost-effective. This is to be encouraged, although one must be careful that, in considering the detail, major choices are not missed.

However, the consideration of institutional care costs is typically much less detailed. Certainly there is never any consideration of whether such care is being delivered in the most cost-effective way. Of greater concern is the fact that most authors confine their analysis to average operating costs. There is therefore the possibility that these costs do not reflect the costs of care of the control group of patients (i.e. those that might alternatively be cared for in the community) since these patients may be less dependent than the average. Conversely, most analysts fail to consider capital costs in the institutional alternative. It may be that expansion in community care will avert future capital costs in the provision of institutional care, or enable some facilities to be closed. Consideration of operating costs alone would understate this potential benefit of community care. It is particularly important that it be considered where *large shifts* in the balance of care are implied by the results of a given study. However, the treatment of capital cost savings needs to be considered in the particular context. In some situations the majority of these will be 'sunk costs' where the opportunities for savings or redeployment are minimal.

Turning to gaps in the range of applications appearing in the literature, it is pleasing to find some examples in the mental illness field. Given its importance in terms of overall health care priorities, it has been relatively underexplored in the past. If health care priorities are any guide, the major remaining gaps appear to be in the mental handicap field and in care of the elderly severely mentally infirm.

Contribution to decision making

Given the methodological weaknesses outlined above, do the studies reviewed here make a useful contribution to decision making? First, although some of the costs and benefits are not measured very accurately, a contribution of many of the studies is that they encourage a broad, societal view of the choice between locations for care, beyond the boundaries of any one health care agency. Secondly, despite the estimation problems, the studies, particularly when taken as a group, suggest that community care does represent a feasible, acceptable and lower cost option in many situations, although more research is undoubtedly required.

Nevertheless, there are a number of additional issues, currently overlooked by many (although not all) of the studies, that would be of concern to decision makers in this field. First, there is the concern that adoption of community-orientated care may represent a financial add-on, rather than a cost saving, if such care fails to substitute for institutional care, or if any freed institutional-based resources are used for the care of other patients. (Weissert *et al.* (1980) discuss the former situation in the context of homemaker services, Evans and Robinson (1980) the latter in the context of day care surgery.) Not all studies help decision makers to think through these issues adequately.

Secondly, few studies explore aspects of *implementation* of strategies involving a shift towards community care. For example, there may be resource costs involved in the closure of hospitals or other institutions. Likewise there may be costs in explaining to the community why they should accept (indeed, welcome) community hostels for the mentally ill or mentally handicapped. Finally, there has been little study of alternative mechanisms for achieving such a shift in the balance of care, including options in the *rate or extent* of run-down of large institutions. For example, are some rates of change more acceptable and of lower cost than others; what happens to the staffing levels and cost of a long-term care institution if the least dependent patients are moved into community care?

Certainly there are many more issues that could be explored, and it is particularly encouraging to note that in the United Kingdom (at least) the cost–benefit approach is now being applied routinely in the appraisal of options where at least one of these is a large capital scheme (DHSS 1981). This may be the appropriate forum within which to explore some of the issues of implementation outlined above and where bridges could be built between health service decision making and the existing studies, many of which are carried out within a research setting.

References

Department of Health and Social Security (1981). *Health Services Management: health building procedures.* HN(81) 30. DHSS, London.

Drummond, M. F. (1980). *Principles of economic appraisal in health care,* Oxford University Press.

Williams, A. H. (1981). Welfare economics and health status measurement. In *Health, economics and health economics* (ed. J. van der Gaag and M. Perlman). North Holland, Amsterdam.

5 Bloom, B. S. and Kissick, P. D. (1980). Home and hospital cost of terminal illness. *Medical Care* **18**(5), 560–4.

1. Study design

1.1. *Study question*

What are the comparative costs of the care of terminally ill patients that die at home or in hospital?(a)

1.2. *Alternatives appraised*

Home care *versus* (acute) hospital care (for the last two weeks of life for patients dying of a malignant disease).

1.3. *Comments*

2. Assessment of costs and benefits

2.1. *Enumeration*

The costs considered were the billed charges under each form of care. These included laboratory tests, radiology, physical therapy, respiratory therapy, radiation treatment, nursing, drugs, supplies, equipment, home-maker services, and physician care. For the home care option, loss of family income was also considered.

2.2. *Measurement*

Since the comparison was not based on a randomized controlled trial, the key measurement issue revolved around the assembly of two groups: matched for age, sex, marital status, occupational category, disease, and site of malignancy. (Matching by disease complications proved too difficult.) In all, 19 pairs of patients were studied retrospectively.

2.3. *Explicit valuation*

Market prices were used to value the cost items identified above. In addition, some anecdotal material concerning the family's reaction to the process of caring for their family member was also recorded.

3. Allowance for differential timing and uncertainty

Discounting was not particularly relevant in this case. No sensitivity analysis was performed but, given the variation in costs by patient, statistical tests of significance were performed.

4. Results and conclusions

It was concluded that home care is economically an important alternative to hospital care for terminal illness. On average, hospital costs were 10.5 times higher, varying from 4 to 73.7 times higher depending on cancer site. A major source of the higher hospital costs was the more intensive use of diagnostic and therapeutic technology, even though it was clear that the patient was going to die. In addition there were large differences in the cost of palliation: US$70 for those at home compared to US$1763 for those in hospital. Although there may be differences between the two patient groups, the authors indicated this as being one area for further study.

5. General comments

The authors noted that, to some extent, the cost differences are to be expected and that a priority area for further research would be the qualitative aspects of care in each setting.

Loss of income by family members accounted for 13.3 per cent of home care costs. Two issues arose: (*i*) would it make sense to compensate family members for nursing sick relatives at home? (*ii*) since family members make this choice freely can we assume that the utility to them exceeds any costs they incur? (On this latter point see the summary of the article in this section by Evans and Robinson 1980.)

Chamberlain, A. (1980). The estimation of costs and effectiveness of community-based family planning services, *International Journal of Social Economics* **7**(5), 260–72.

1. Study design

1.1. *Study question*

What is the cost of family planning (FP) service provision and how effective are these services?(c)

1.2. *Alternatives appraised*

FP services *versus* 'other sources', such as retail pharmacies (which would be used if FP services were not available).

1.3. *Comments*

2. Assessment of costs and benefits

2.1. *Enumeration*

The costs considered were those falling on the health service for the provision of community-based FP services. They included those of specialist FP clinics, sessions in multi-purpose health centres and clinics, GP costs and the cost of the contraceptives. Costs to clients were not included. Effectiveness was considered in terms of avoidance of unwanted pregnancies. Other benefits of the FP services were mentioned, but not included in the study; for example, screening for cervical cancer and administration of rubella vaccine.

2.2. *Measurement*

The effectiveness of the FP services was measured as a function of the number of 'at risk' females recruited and the increased effectiveness of the FP services when compared with contraceptive techniques available in the absence of this service. In carrying out the calculations, it was assumed that one contraceptive method, oral contraceptives, was used for the FP service clients. In the absence of FP services, it was assumed that the most effective alternative, the sheath, would be used. The utilization rate for FP services was calculated after allowance had been made for the prevalence of sterilizations and the number of women attempting to become pregnant, or already pregnant.

The measurement of costs attempted to identify marginal costs for multi-purpose clinics. Only a small proportion of costs could not be appropriately allocated.

2.3. *Explicit valuation*

Estimates of use–effectiveness for the different contraceptive methods were taken from the literature. The value of the reduction in unwanted pregnancies was not assessed since this was not required by the study question. The costs were based on wages and prices in the NHS, i.e. market prices.

3. Allowance for differential timing and uncertainty

Different estimates of use–effectiveness were used to give upper and lower limits for the effectiveness of FP services. Differential timing was not relevant to the study.

4. Results and conclusions

The upper and lower estimates for avoided conceptions were 1242 and 828. Similar estimates for avoided births, after allowing for spontaneous but not therapeutic abortions, were 1012 and 675. The total cost for FP services in 1977 was estimated at £328 000. Therefore, the cost per pregnancy averted was between £264 and £396, and the cost per unwanted birth avoided was between £324 and £486. The author concluded that these cost figures 'seem low when compared to the magnitude of the costs which would have been incurred by society and the female (and her family) had the unwanted pregnancy gone to term'.

5. General comments

The paper also reported estimates of the relative costs of alternative sources of provision. These were £10.74 per attender per annum for GP services, £10.22 for a specialist FP clinic, and £7.87 for provision in multi-purpose clinics. The author stressed that these were average costs and sensitive to throughput rates.

7 Doherty, N., Horowitz, P. A., and Crakes, G. (1980). Real costs of dental care in private and public practices. *Medical Care* **18**(1), 96–109.

1. Study design

1.1. *Study question*

What accounts for the differences in costs per patient in three modes of delivering dental care?(a)

To what extent can the observed variations in normal costs under these modes of delivery be explained economically by differences in outputs, production processes, and inputs?

1.2. *Alternatives appraised*

Mobile units *versus* private practice *versus* public practice. (See 1.3 below.)

1.3. *Comments*

The main purpose of the study was not to estimate the relative costs in the three modes, but to see whether these could be explained by different production processes. Earlier findings had indicated cost differences which could not be explained by variations in socio-demographic characteristics in children, or quality and quantity of dental care.

The study was based on the Chattanooga Project, a publicly-funded programme providing dental care to indigent children by private practice and public fixed and mobile clinics.

2. Assessment of costs and benefits

2.1. *Enumeration*

The study only considered the direct costs of providing the services, and did not include costs to the patients.

2.2. *Measurement*

The total costs of each mode of care to the Project were calculated, and taking the number of patients and patient visits, and chair time for each mode, the authors arrived at the average costs per patient, per patient visit, and per hour of chair time for the three modes of delivery.

In analysing the reasons for the cost differences the study did not develop relationships between costs and any particular factor, but

Doherty *et al.* (1980)

analysed cost differences among modes, factor by factor. This was achieved by use of a three factor regression model (output, productivity, and input), and then using a stepwise procedure which adjusted costs to eliminate the effects of (*i*) different outputs, and (*ii*) differing productivity rates, as possible causes of the cost differences. The adjusted set of costs reflected equality across the modes in both outputs and productivity. Cost differences that still remained would therefore be insignificant if there were only small variations in returns to the third component, the inputs.

2.3. *Explicit valuation*

The costs were collected during an earlier study (see 5 below), and therefore full details were not given. Costs for dentists in the private sector were reflected by their fees, and in the public sector by their salary. It is assumed that market values were used for capital equipment, supplies, services, and utilities, although this is not stated in the paper.

3. Allowance for differential timing and uncertainty

Differential timing was not particularly relevant in the context of this study apart from the treatment of capital costs, the nature of which differ between practice modes. The authors stated that, in their earlier study, 'depreciation and interest on buildings was considered' but details are not given.

The authors pointed out a number of uncertainties, some relating to the regression model itself and others relating to the limited data base. No sensitivity analysis was performed, however.

4. Results and conclusions

Private costs were significantly higher than public costs, but there was no significant differences between the public modes. The productivity and service adjustments eliminated the differences in costs between the public modes. On all bases, real private mode average costs were 66 per cent higher than fixed public mode costs and 87 per cent higher than mobile mode costs. These differences were due to differences in inputs, and dental resources earn significantly higher returns per patient in the private than in the public mode. The authors suggested that significant efficiencies may well be gained by more appropriate use of public practices.

212

5. General comments

The authors used a complex regression model, which is explained in an appendix.

Further details of the costing procedures used are given in Doherty, N. and Vivian, S. (1976). Cost of publicly financed dental care for children in three different types of practice settings. *Journal of Public Health Dentistry* **36**, 3–8. (This paper is summarized in the first volume of *Studies*, 1981.)

As was stated in the earlier summary, comparisons of the efficiency of care in different modes require some explicit consideration of the quality of care given. Nevertheless, the results are sufficient, as the authors suggested, to question the accepted wisdom that, in comparison with private practices, public practices are inefficient and provide lower quality care. (This is the basis of policies of encouraging, through subsidization schemes, the greater use of private practices.) However, further study of accessibility and quality of practice modes is probably required.

68 Evans, R. G. and Robinson, G. C. (1980). Surgical day care: measurements of the economic payoff. *Canadian Medical Association Journal* **123**, 873–80.

1. Study design

1.1. *Study question*

By how much are the treatment costs at a children's hospital reduced when patients are cared for in a surgical day care unit rather than in an inpatient ward?(a) (See 1.3 below.)

1.2. *Alternatives appraised*

Surgical day care (patients admitted and discharged on the same day) *versus* inpatient care (for a range of dental and non-dental diagnoses).

1.3. *Comments*

One of the major purposes of this paper was to point out the methodological deficiencies in earlier approaches to the estimation of cost savings, based on *per diem* charges.

2. Assessment of costs and benefits

2.1. *Enumeration*

The costs considered were those arising from hospital resources consumed in each alternative. The relevance of patient and family costs was mentioned but it was argued that, since patients and families had a free choice of either form of care, the relative benefits to them of surgical day care were at least equal to any additional time and money costs borne by them. (See 5 below.)

2.2. *Measurement*

The study was based on a prospective analysis of two cohorts of patients—2169 undergoing surgical day care, and 460 'quasi-controls' judged appropriate for surgical day care. The authors acknowledged the possibility of other designs, such as random allocation of patients to the regimens but argued that, at least for the consideration of costs, experimental activity may have induced changes in the care delivery system thereby producing results atypical of a regularly operating system.

214

2.3. *Explicit valuation*

The units of service consumed by patients from direct service and indirect service (overhead) departments were calculated. These units were then combined with hospital operating expenditures by department to obtain treatment costs.

3. Allowance for differential timing and uncertainty

Discounting was not relevant in the context of this study.

A sensitivity analysis of inpatient care costs was performed, by reducing the observed average length of stay (3.79 days) to its previous year's value (2.7 days) and by reducing the observed utilization of diagnostic procedures to the lower level observed for day care patients. The authors argue that taken together these adjustments would produce a 'minimal' estimate of inpatient care and thereby a conservative estimate of the savings from surgical day care.

4. Results and conclusions

The cost analysis showed that there were *potentially* very large savings from the substitution of surgical day care for inpatient care. However, the extent to which these savings would be realized depended on the behavioural responses of the hospital. For example, would the hospital adjust its staffing when bed use falls? Would the development of surgical day care result in the admission of other inpatients?

The authors note that the potential savings had not been achieved and go on to discuss the ways in which incentives could be changed (within the Canadian system) in order to realize such savings.

5. General comments

The authors argued that in this case it was not necessary to undertake a survey of patient and family costs of day care. The reasoning for this is that, at least *ex ante*, day care was a preferred option for those who chose it. However, it still may be of interest to examine patient and family views *ex post*. In this case supportive evidence was available from an earlier study in the same hospital. Other studies present a slightly 'muddier' picture of patients' preferences for shorter hospital stays. See, for example, Waller *et al.* (1978). *Early discharge from hospital for patients with hernia or varicose veins*. Department of Health and Social Security, HMSO, London: summarized in the first volume of *Studies*, 1981.

69 Fenton, F. R., Tessier, L., Contandriopoulos, A.-P., Nguyen, H., and Struening, E. L. (1982). A comparative trial of home and hospital psychiatry treatment: financial costs. *Canadian Journal of Psychiatry* **27**(3), 177–87.

1. Study design

1.1. *Study question*

What are the comparative costs of home and hospital treatment for psychiatric patients?(a) (See 1.3 below.)

1.2. *Alternatives appraised*

Community-based treatment (outpatient care plus home treatment) *versus* hospital-based treatment (having inpatient treatment as a major constituent). (See 1.3 below.)

1.3. *Comments*

Although this paper restricted itself to costs, the work reported is part of a broader study in which the medical effectiveness of the alternatives was assessed by a randomized, controlled trial. The study group comprised patients for whom the diagnosis was schizophrenia (40.6 per cent), manic-depressive psychosis (29.1 per cent), and depressive neurosis (30.3 per cent).

2. Assessment of costs and benefits

2.1. *Enumeration*

The study considered both the costs of the treatment itself and the costs of consequences of illness during the year of study. Costs of treatment included manpower and operating costs, drug costs defrayed by patient or government, psychiatrist's fees, and transportation costs (defrayed by family). The costs of consequences of illness included loss of working days and salary, cost of housekeeper (defrayed by family and by social welfare), and money given or lent to patient by family.

2.2. *Measurement*

Only manpower and operating costs were measured in money terms (see 2.3 below) and were obtained from financial records and patient records. Other items of cost were expressed in terms of whether they were present or absent for each patient, with the exception of number of working days lost by the patient and family (which was expressed in numerical terms). Costs of treatment *excluded* laboratory, radiologic, food, psychotropic drugs, and capital costs. In addition the costs of treatment received by patients and families in social and vocational rehabilitation programmes outside the hospital were excluded.

2.3. *Explicit valuation*

Health service expenditures were used to estimate operating and manpower costs of the treatments. Hospital care was costed both at a *per diem* rate and at differential daily rates which reflected the intensity of different types of care within the hospital.

3. Allowance for differential timing and uncertainty

No allowance was made for differential timing of costs as only a one year time-span was considered.

Cost estimates were presented using both the average *per diem* figure (Canadian$143) and the differential daily figures reflecting the intensity of care (Canadian$210 for intensive care, Canadian$40 for regular care).

No other sensitivity analysis was performed, although tests of statistical significance were performed on the differences between costs for the two treatment groups.

4. Results and conclusions

Irrespective of the way of estimating manpower and operating costs, hospital-based treatment was more expensive during the year. (For example, $3250 *versus* $1980 per patient using the costing method which reflected different intensity of care by day.) With two exceptions during the first month of treatment, the proportions of patients and families receiving either treatment who incurred other costs of treatment were low, and the differences between groups were not significantly different. A higher proportion of patients and families receiving home-based treatment defrayed the cost of the patient's psychotropic drugs; secondly, a higher proportion of families of patients receiving hospital-based treatment defrayed transportation costs.

5. General comments

The paper also contains a useful discussion which compares the results obtained with those obtained in other studies.

70 Gibbins, F. J., Lee, M., Davison, P. R., O'Sullivan, P., Hutchinson, M., Murphy, D. R., and Ugwu, C. N. (1982). Augmented home nursing as an alternative to hospital care for chronic elderly invalids. *British Medical Journal* **284**, 330–3.

1. Study design

1.1. *Study question*
What are the costs and effectiveness of augmented home nursing care?(c)

1.2. *Alternatives appraised*
Augmented home nursing care for elderly invalids *versus* geriatric hospital inpatient care.

1.3. *Comments*
In costing the home nursing alternative differentiation was made between those patients living alone in sheltered housing, those living alone in general housing, those living with a spouse, and those living with other relatives.

2. Assessment of costs and benefits

2.1. *Enumeration*
Costs included in the community option comprised personal consumption (heating, food, etc.), health and local authority services (district nursing, home-helps, drugs and dressings, laundry and meals on wheels, plus attendance allowances where applicable). Housing costs were included for patients living alone since, if the people were transferred to permanent long stay institutional care, their housing could be made available for use by others.

GP and chiropody costs were excluded as these were thought to be small. Costs incurred by patients, family and friends were also excluded.

Effectiveness of the scheme was measured in terms of patients' mental and physical functioning and their level of medication.

2.2. *Measurement*
The study was not linked to a prospective clinical trial. Cost and effectiveness measures were based upon the experience of 24 patients admitted to the scheme during a six month period, compared to that of a 'control' group of similar age distribution and principal diagnosis

218

admitted to the geriatric wards of a District General Hospital. The cost of inpatient care for these patients was proxied by the average cost of long stay accommodation in 1980.

In the case of community care costs, personal consumption was based on family expenditure survey figures for households of the relevant income and composition. Service costs were measured directly, with some supplementary information being drawn from Personal and Social Services statistics.

Effectiveness was measured by means of an instrument used for assessing suitability for admission to homes for the elderly. This produced functional rating scores which ranked patients in terms of physical and mental wellbeing and level of medication.

2.3. *Explicit valuation*

Market values were used to estimate costs in most instances. Health and community service costs were based on actual expenditure which, in the case of certain items, might be biased downwards owing to the existence of subsidies. A weekly housing cost was imputed for those patients living alone. For those in general accommodation, this was based on the cost of new buildings and, for those in sheltered housing, on the cost of such new schemes in the given locality, with an allowance for warden costs.

The estimation of the weekly cost of a hospital bed was based on DHSS estimates (these are average figures, including revenue (running) costs only).

3. Allowance for differential timing and uncertainty

Capital costs of housing and hospital bed provision were considered and converted to a weekly charge. The discount rate applied in the case of the former was not stated, but in the case of the latter was the UK Treasury test discount rate.

4. Results and conclusions

Of those admitted to the home nursing scheme who were assessed twice, none showed any decrease in functional ability over the six month period and some showed a small improvement.

There was considerable variation in the weekly cost per case cared for at home. Of the 21 cases costed in detail, the range of cost per week varied from £46 to £194. The average weekly cost for those living alone in sheltered accommodation was £126; for those living alone in general accommodation it was £91; for those living with a spouse it was £118 and for those patients living with other relatives it was £101.

As a group, only those living alone in sheltered accommodation incurred a higher average weekly cost than the average revenue costs of long stay care. When capital costs of inpatient care were added to the revenue costs, then only one of the patients treated at home incurred a higher cost.

5. General comments

The small study population, combined with the methodology employed in the study, means that few firm conclusions could be drawn.

Changes in the functional status of the patients cared for at home was measured for only nine patients. Furthermore, of the 24 patients allocated to home nursing care, only six had been in the scheme for more than five months. Changes in the functional status of the hospital 'control' group were not recorded.

The criteria for allocation to home nursing care were not set out, but of 46 patients referred for such care only 24 joined the scheme. On the basis of the functional measurement instrument used, it appeared that, by-and-large, those patients accepted for home care had a lower level of physical and mental functioning than did those who were rejected; but higher levels than those treated in hospital. Therefore, the control and study groups may well not have been alike in all respects.

Costs falling on patients' families were ignored. This would seem an important omission, especially as the authors noted that support from family or friends affected the level of care provided to patients by the outside agencies.

One of the main justifications for the augmented home nursing scheme (made by the authors) was that a reduction in the number of additional long stay geriatric beds would be possible. In estimating the cost consequences for the health service of such a change, the costs of community care provision should have been compared with the *marginal* rather than the *average* cost of inpatient care. That is, the average costs of current care may overstate the potential savings. (See the companion volume, Section 3.2.)

71 Gray, A. M. and Steele, R. (1981). The economics of specialist and general practitioner maternity units. *Journal of the Royal College of General Practitioners* **31**, 586–92.

1. Study design

1.1. *Study question*

What are the comparative costs of specialist and GP maternity units in terms of cost per delivery, cost per occupied bed-day, and cost of antenatal, labour/delivery and postnatal care?(a)

1.2. *Alternatives appraised*

Maternity care provided in specialist units *versus* such care provided by GP units.

1.3. *Comments*

The authors acknowledged that the two types of provision of maternity care were not viewed as perfect substitutes for all patients; the specialist units generally treated higher-risk mothers and GP units were typically allocated low-risk mothers with expectations of a normal delivery. In addition, in the Health Board under study, a policy of 48-hour postnatal discharge from specialist to GP units was in operation and to this extent GP units were acting as a complement to, rather than a substitute for, specialist units.

2. Assessment of costs and benefits

2.1. *Enumeration*

Only NHS expenditure was considered; the authors acknowledged that family and other social costs were important, but often unquantifiable. Consideration was limited to the running costs of the two types of maternity care as capital costs were deemed to be outside the scope of the study.

2.2. *Measurement*

The costs reported in the study were derived from the authors' own survey data.

The authors sought to determine the amount of expenditure, in mixed speciality GP units, which was spent upon obstetric care, and the proportion of expenditure in both GP and specialist maternity units which was allocated to antenatal, labour/delivery, and postnatal care.

221

Gray and Steele (1981)

2.3. *Explicit valuation*

The method of valuing costs was not made explicit; presumably market prices were used. Costs were expressed in 1976/7 prices.

3. Allowance for differential timing and uncertainty

Differential timing was not relevant in the context of this study. No sensitivity analysis was performed.

4. Results and conclusions

The average cost per delivery in a GP unit was £393 compared to £484 in specialist units. Costs per in-patient day were £47 and £52 respectively.

In all three phases of care (antenatal, labour/delivery, and postnatal), GP units were found to cost less than specialist units on a *per diem* basis. The average variable cost in GP units per day of antenatal care was 51 per cent of the corresponding costs in a specialist unit; the cost per labour/delivery day in the GP units was 32 per cent of that in the specialist unit, and the cost of postnatal care in a GP unit was 59 per cent of the corresponding *per diem* cost of the specialist unit.

Expenditure in GP units was heavily weighted towards the provision of postnatal care, which accounted for almost 70 per cent of all inpatient expenditure. Expenditure in specialist units, conversely, was heavily weighted to the other two phases of care. This difference was mainly explicable in terms of the 48-hour discharge policy which was in operation from the specialist to GP units. Given the cost differential between GP and specialist units, the authors felt that the 48-hour discharge policy would be cost-effective.

The authors tested the hypothesis that the average cost per delivery in GP units could be decreased by increasing occupancy and found an inverse, but weak correlation (0.4) between the average cost per GP unit delivery and the occupancy rate. This suggested that as occupancy rates increased, the average cost per delivery would decrease, but not by a very large amount.

5. General comments

Details of the costing methodology employed are reported more fully in Gray, A. M. and Steele, R. (1979). The identification of the costs of maternity care: a programme approach to health service expenditure. *HERU Discussion Paper No. 03/79*. Department of Community Medicine, University of Aberdeen.

222

Kriedel, T. (1980). Cost–benefit analysis of epilepsy clinics. *Social Science and Medicine* **14C**, 35–9.

1. Study design

1.1. *Study question*
Is the provision of epilepsy clinics worthwhile?(d)

1.2. *Alternatives appraised*
System of epilepsy clinics *versus* (implicitly) maintaining the *status quo*.

1.3. *Comments*

2. Assessment of costs and benefits

2.1. *Enumeration*
The costs considered were those relating to the provision of 46 adult and 46 child clinics and operating them over a period of 30 years. The benefits comprised cost savings from a reduction in the need for institutional care and the gain in health status for the patients. Other costs and benefits to the patients and their families were not considered.

2.2. *Measurement*
A feature of the study was the construction of a health status index to measure changes in health outcomes (see Appendix 3 of the companion volume). Five functional levels were used; 1 = fully active, 5 = dead. The probabilities of moving between levels were taken from existing literature and incorporated into a mathematical model. The relative values of the different function levels were derived from a small sample of the general population (54 respondents). With estimates of the projected incidence of different forms of epilepsy, health status gains over the 30 year life of the programme were calculated and expressed as 'function years' gained from the provision of clinics.

2.3. *Explicit valuation*
The function years gained were valued by average wage rates representing a minimum estimate of the value of time. Market values were used to estimate costs.

Kriedel (1980)

3. Allowance for differential timing and uncertainty

Health status gains were discounted at 3 per cent per annum. Resource costs and benefits were discounted at 7 per cent, but there was no discussion of the rationale underlying the use of different discount rates.

Although a number of assumptions were made in the analysis and the values used in the construction of the health status index came from a small sample, there was no attempt to test the sensitivity of the results.

4. Results and conclusions

It was concluded that '... each patient treated by the epilepsy clinics will experience an improvement in his health status that the individual appreciates as equal to two years of health'.

Assessed over a period of 30 years, the discounted costs of establishing and running the clinics was DM0.53 billion, whilst the present value of health status gains and cost savings was DM2.79 billion. The author argued that 'these results demonstrate that the epilepsy program is efficient and should be realized as soon as possible'.

5. General comments

73 Linn, B. S., Stephenson, S. E., Bergstresser, P. and Smith, J. (1979). Do dollars spent relate to outcomes in burn care? *Medical Care* **17**(8), 835–43.

1. Study design

1.1. *Study question*

Does additional expenditure on the treatment of burns improve outcome?(c)

1.2. *Alternatives appraised*

Treatment in hospitals with a special burn unit *versus* treatment in hospitals without special facilities.

1.3. *Comments*

The study was a retrospective analysis of new patients admitted with burns in 75 Florida hospitals during a one year period. There were 252 patients admitted to hospitals with specialist burn units and 1012 admitted to hospitals without such facilities.

2. Assessment of costs and benefits

2.1. *Enumeration*

The study compared the average treatment cost with the patient outcome in terms of survival (mortality), complications (morbidity), and length of stay for two groups of patients. Only hospital costs were considered; costs for other health care services and costs falling on the patient and family were ignored, although different lengths of stay might have implications for these. Other measures of outcome, such as time of return to work, could not be included because of the retrospective nature of the study.

2.2. *Measurement*

Data on morbidity, mortality and length of stay were taken from patient medical records. Morbidity was measured by the number of patients with complications in five categories; burn shock, pulmonary, wound sepsis, psychiatric, and contractures. No account was taken of the severity of complications, but the severity of the burn was recorded.

Treatment costs were proxied by the hospitals' charges to the patients. Not only is this an average, rather than a marginal cost, but it must also

225

be noted that there may be variations between the hospitals in the charge made for the same resource use. Actual resource use by the two patient groups is not reported.

2.3. *Explicit valuation*

As noted above, costs were determined by prevailing hospital charges. Outcome measures were not valued or weighted in any way.

3. Allowance for differential timing and uncertainty

Differential timing was not relevant in this study.

The patient data were analysed in three ways. First, all burn patients in the two types of hospital were compared after allowance was made for differences in age and the severity of burn between two groups. Secondly, the comparison was restricted to patients with severe burns (over 30 per cent body surface area burned), also adjusted for age and severity. Finally, the analysis was limited to patients for whom controls could be obtained, matched for age *and* severity of burn.

4. Results and conclusions

The first comparison showed that the patients treated in hospitals with special burn care facilities had higher average charges, US$2849 against US$1540, and also experienced significantly greater morbidity and mortality. They also spent seven days longer in hospital on average. A similar pattern emerged when the comparison was restricted to severe cases, although fewer of the results reached statistical significance. For the comparison of matched patients, the numbers were reduced to 39 in each group, making it more difficult to achieve statistical significance. The charges were still significantly higher for patients in the special burn units, US$4967 against US$3985, but only one measure of morbidity was significantly worse for these patients. The other measures of outcome were generally worse for the special burn unit group but were not statistically significant. The authors concluded that the results suggest that additional expenditure on burn patients does not improve outcome but, because of the limitations of the data used, they argued that a prospective randomized controlled trial is needed to confirm the results.

5. General comments

The authors discussed at some length the difficulties inherent in relying on retrospective data from medical records for a study of this type. That is, there is likely to be a selection bias in referral of patients to hospitals with specialized burn units or *vice versa*. Therefore one cannot be confident that the patients in the two types of institution are alike in all respects. However, the authors offered no comment on the costing used in the study. The distinction between resource use and hospital charges was ignored but this is crucial. As the data stand, it is impossible to detect whether the results of the study were, to some extent, the product of different charging policies.

74 Logan, A. G., Milne, B. J., Achber, C., Campbell, W. P., and Haynes, R. B. (1981). Cost-effectiveness of a worksite hypertension treatment program. *Hypertension* **3**(2), 211–18.

1. Study design

1.1. *Study question*

Is it more cost effective to treat hypertension at the patient's place of work or in the community?(c)

1.2. *Alternatives appraised*

Treatment at the worksite *versus* treatment in the community from physicians in private practice.

1.3. *Comments*

Participants for the study were selected from 21 906 volunteers aged 18 to 69 years in 41 business locations in metropolitan Toronto who were screened for hypertension in 1976–77. Worksite care was provided by nurses working to a standard protocol. In the community care approach an appointment was made with the individual's own doctor.

2. Assessment of costs and benefits

2.1. *Enumeration*

The costs considered were those falling on the health care system and the patient as a result of screening and treatment. These included (for screening) personnel, equipment, supplies, travel, participants' time, and administrative costs, and (for treatment) provision of care, laboratory examinations, hospitalization, and drugs. The effectiveness of the treatment was assessed in terms of blood pressure (BP) reduction.

2.2. *Measurement*

The study was based on a randomized controlled trial. Individuals admitted to the trial were followed up at six and 12 months. BP was measured and a questionnaire administered to determine medication status.

2.3. *Explicit valuation*

Market prices were used to value costs; physician fee schedules were used to estimate medical personnel costs; participants' time was valued according to hourly wages. Administrative costs (for the screening programme) were estimated at 30 per cent of health system costs. Hospitalization costs were estimated using *per diem* rates. (More elaborate methods were rejected because of the small numbers involved.) The patient's cost of hospitalization was taken as the monetary value of time lost from work.

3. Allowance for differential timing and uncertainty

No discounting of future costs and effects was employed because of the short duration of the study (one year). A sensitivity analysis was carried out to correct for any biases caused by missing cost data. This was performed by substituting (for missing data on patients' drug costs and travel times) the *highest* individual cost in the worksite group and the *lowest* in the community care group. This had the effect of underestimating any cost advantages of the worksite programme compared to the result that would be obtained by substituting average cost data for the missing data in both groups.

4. Results and conclusions

The average total cost per patient for worksite care over the 12 month period was Canadian $243, compared with $211 for regular community care. (This difference was not statistically significant.) The worksite *health system* cost was significantly more expensive ($197 *versus* $129) but the *patient* cost was significantly lower ($46 *versus* $82). The worksite programme was more effective, however, in that the *incremental* cost effectiveness ratio (reflecting the returns from the extra resources used in worksite care) was $5.63 per mmHg compared to the base cost-effectiveness ratio of $32.51 per mmHg for regular care. The authors argued that the findings support health policies that favour allocating resources to work-based hypertension treatment programmes for the target group identified in the study.

5. General comments

This article is one of few to conduct statistical tests of significance of observed cost differences. The basis for cost comparisons were the real data obtained, using the sensitivity analysis to correct for biases arising from missing data. There are other reasons for conducting sensitivity analyses, however. For example, the costs or effects of the innovative programme may not be the same in another setting employing less enthusiastic personnel. (See the companion volume, Chapter 5.) The study also makes it clear that it is the *incremental* cost effectiveness ratio that is important when making programme comparisons.

75 MacFarlane, J. P. R., Collings, T., Graham, K., and MacIntosh, J. C. (1979). Day hospitals in modern clinical practice—cost benefit. *Age and Ageing* **8**, Supplement, 80–86.

1. Study design

1.1. *Study question*

To what extent does the day hospital achieve its objective of rehabilitation; what are the costs of day hospital and supporting community care; how do these compare with inpatient costs? (b) and (a) (See 1.3 below.)

1.2. *Alternatives appraised*

Day hospital care plus community support for the elderly *versus* inpatient care.

1.3. *Comments*

The study comprised mainly a survey of day hospital practices, although a simple costing exercise was presented. The authors did not claim that the study represented a full cost-effectiveness analysis of day hospital care *versus* inpatient care.

2. Assessment of costs and benefits

2.1. *Enumeration*

Benefits were measured against criteria internally set. The study posed the questions; where do the patients come from, what is wrong with them, what is done for them, and what is their status at the end of treatment?

Total costs of the day hospital were calculated and expressed on the following bases: per annum, per week, and per patient visit (assuming 82.5 per cent occupancy). The total weekly cost of day hospital care was calculated by including the cost of community support, and was compared to inpatient costs by week, and episode of care.

2.2. *Measurement*

The study was not linked to a controlled clinical trial. The benefits of day hospital attendance were assessed by consensus of opinion from the doctor, the nursing staff, the therapists, the social worker, and all who had participated in treatment. Day hospital costs were broken down into costs of salaries, laboratory tests, drugs, general services, catering, and ambulance services. The cost per patient visit was derived by taking the total cost per patient week, and dividing by the average number of visits per week (2.5). Neither the costs of weekly community support services, nor of the weekly inpatient costs were similarly broken down. All costs used were average costs.

2.3. *Explicit valuation*

It is assumed that market prices were used. It is not clear whether employer's overhead costs were included in the estimates of salaries. Capital costs appeared to be omitted from the day hospital cost estimates. Capital costs of ambulances were explicitly mentioned, but were not correctly converted to an equivalent annual cost using a discount rate. (See Chapter 2 of this volume.)

3. Allowance for differential timing and uncertainty

Not considered. Differential timing was not particularly relevant in the context of this study.

4. Results and conclusions

The survey of benefits confirmed that rehabilitation was still the main objective and achievement of the day hospital. The average number of reasons for attending the day hospital was five (a smaller number of reasons (one or two) would imply a more economic placing as an out-patient). More than one-third of patients discharged were considered to have improved, and a further one-quarter were maintained in the face of advancing disease; therefore a majority of patients were considered to have benefited from day hospital attendance.

Total day hospital cost was £2254 per week or £13.60 per patient visit. Community support was taken to be £13 per week, and therefore the weekly cost (per patient) of day hospital and community care was calculated to be £47, compared to an inpatient cost of £129.80p per week. The total cost of an average 13-week course of day hospital treatment was calculated to be £611, and the average 103-day inpatient stay in hospital to be £1910. The authors therefore concluded that 'the day hospital does provide a genuine saving in money while providing an effective therapeutic service'.

5. General comments

This was a simple study, and a number of improvements in the

MacFarlane *et al.* (1979)

measurement of benefits and costs could be made. A controlled comparison of the benefits of day hospital care *versus* inpatient care would be advisable. Some of the cost estimates are open to question; is it unclear whether capital costs are included, or are discounted, how the costs are derived or which costs are included. The use of average costs, particularly for inpatient care, is again a problem. It is not clear whether money will be saved by placing the elderly in day hospitals, since these savings may not be realizable. It is also not clear for whom day hospital treatment is more appropriate.

6 Mangen, S. P., Paykel, E. S., Griffith, J. H., Burchell, A., and Mancini, P. (1983). Cost-effectiveness of community psychiatric nurse or out-patient psychiatrist care of neurotic patients. *Psychological Medicine* **13**, 407–16.

1. Study design

1.1. *Study question*

What is the most cost-effective way of providing follow-up care for neurotic outpatients?(c)

1.2. *Alternatives appraised*

Care by community psychiatric nurses (CPN) as the main therapist *versus* routine outpatient psychiatrist follow-up.

1.3. *Comments*

The cost-effectiveness study was the third in a series dealing with this topic. Earlier papers deal with clinical and social aspects of care. (See 5 below.) The study sample comprised patients aged 18–69 years who had either been discharged from hospital or day care, or who were outpatients already attending for six months. To qualify, patients had to be considered by their psychiatrist to require at least six months further follow-up care. Diagnostically, the main criterion was a primary diagnosis of neurosis, affective psychosis, or affective, schizoid, anankastic, hysterical or asthenic personality disorders.

2. Assessment of costs and benefits

2.1. *Enumeration*

The costs considered included use of all psychiatric treatment resources, non-psychiatric treatment resources, general practitioner resources, local authority welfare services, and patient travel costs incurred in seeking treatment. In addition, transfer payments (e.g. supplementary, unemployment or sickness benefits) were considered as the authors were interested in investigating the overall impact on public expenditure. (See 5 below.) The effects considered included symptom alleviation, social role performance, family burden and effect on relatives' work record, consumer satisfaction, effects on work records of patients in attending for treatment, and travel time to attend for treatment.

Mangen *et al.* (1983)

2.2. *Measurement*

The study was based on a prospective controlled trial. Patients were randomly assigned to CPN care (n = 35) or regular outpatient care (n = 36), and followed-up for 18 months. Assessments were made at three six-monthly intervals. These included assessments of the non-monetary effects and of usage of psychiatric and non-psychiatric services. These later assessments were then combined with unit cost data to estimate the total costs of each regimen.

2.3. *Explicit valuation*

No attempt was made to place monetary values on the clinical or social effects of therapy. NHS, local authority, and patient expenditure data (i.e. market prices) were used to estimate costs. Information from the accounts of the relevant bodies was used to derive average (unit) costs, with adjustments for overheads made as appropriately as possible. The authors suggested that these average costs 'may approximate to long-run marginal costs (i.e. changes in total cost attributable to a marginal expansion in the provision of service) but they are likely to over-estimate them slightly, since they assume that there is no spare capacity'.

3. Allowance for differential timing and uncertainty

Since the study concerned the current costs of care of patients over a relatively short period, discounting was not relevant. (See 5 below.)

Tests of statistical significance were applied to ascertain whether there were differences in observed costs and effects between the two groups. No formal sensitivity analysis was performed, but the authors discussed in some detail those factors having a great impact on costs (e.g. whether or not a child was placed in care, and whether or not an elderly person attended an elderly day centre). The stability of costs through time was also discussed, with a view to ascertaining whether any financial advantage accruing to CPN follow-up would remain beyond the time period of this study.

4. Results and conclusions

There was no statistically significant difference in the public expenditure for the two modes of care. The costs of psychiatric care were initially greater in the CPN group, but over the whole 18-month study period nursing was a cheaper option in terms of these costs. In any case, the costs of psychiatric care comprised a small proportion of total public expenditure. Clinical and social outcomes were comparable in both follow-up groups but consumer satisfaction was significantly greater among the CPN patients. The authors argued that 'on balance, these results confirm the benefit of community psychiatric nursing for this patient group'.

5. General comments

The clinical and social effects of the regimens are reported more fully elsewhere: Paykel, E. S. *et al.* (1982). Community psychiatric nursing for neurotic patients: a controlled study. *British Journal of Psychiatry* **140**, 573–81; Mangen, S. P. and Griffith, J. H. (1982). Patient satisfaction with community psychiatric nursing: a prospective controlled study. *Journal of Advanced Nursing* **7**, 477–82.

The study considered a mixture of real resource costs and transfer payments. (See the companion volume, Section 2.1.1.) However, the authors clearly specified their aims in doing this and the nature of each monetary effect is clearly identified.

Strictly speaking discounting to present values would be relevant, as the time profile of some costs was different between the options. (For example, the psychiatric costs of the CPN group were higher early on.) However, the *numerical* impact of discounting by a small positive rate (e.g. 5 per cent) would be minimal and the authors were justified in ignoring this.

77 Mowat, I. G. and Morgan, R. T. T. (1982). Peterborough Hospital at home scheme. *British Medical Journal* **284**, 641–3.

1. Study design

1.1. *Study question*

What are the costs and acceptability of a hospital at home scheme for patients who would otherwise have been assigned to inpatient management?(c)

1.2. *Alternatives appraised*

Care administered in the patient's home *versus* inpatient hospital management.

1.3. *Comments*

Criteria for admission of patients to the hospital at home scheme were that: (*i*) had it not been for the scheme the patient would have been admitted to hospital, and (*ii*) the hospital at home staff, patient, patient's relatives, and GP all had to prefer domiciliary care to hospital admission.

The second criterion might be expected to have introduced a selection bias which would effect carers' and patients' evaluations of the scheme.

2. Assessment of costs and benefits

2.1. *Enumeration*

Only the running costs (exclusive of medical and drug costs) of the alternatives were considered. No account was taken of differential costs falling upon patients and their families, nor of any differences in the costs of post-discharge management in the community. The success of the hospital at home scheme was based on the success ratings given by patients, their families, and the health care professionals involved.

2.2. *Measurement*

This study was not linked to a randomized controlled trial.

Costs of the hospital at home scheme were extracted from its expenditure statement for the period April to November 1980. These were compared with Peterborough Health District figures relating to GP, partly acute, and mainly acute hospitals.

Assessment of the acceptability of the scheme was measured via GPs responses to a questionnaire. In the case of district nurses, relatives, and

patients, evaluations were obtained from responses to structured interviews. Evaluations were sought from, or on behalf of, one-quarter of the
200 patients admitted to the scheme during the study period.

GPs involved in evaluating the scheme were asked to identify the type
of inpatient management that was substituted by the hospital at home
alternative.

2.3. Explicit valuation

Market values were used. Costs were expressed at 1980 price levels.
Peterborough hospital costs relating to 1979/80 were inflated by 12.5 per
cent in order to make them comparable with hospital at home costs
relating to 1980/81.

3. Allowance for differential timing and uncertainty

No sensitivity analysis was performed. In particular, there was a wide
range of age and morbidity in the patients admitted to the hospital at
home scheme, but no estimates were provided of how representative the
average costs presented were for these different groups. Differential
timing was not particularly relevant in the context of this study.
However, if it were to be argued that the hospital at home scheme
averted capital expenditure then any capital sums would have to be
annuitized using a discount rate. (See Chapter 2 of this volume.)

4. Results and conclusions

The average age of patients included in the study group was 71 years; the
range being 34 to 95. According to the GPs' views, the scheme provided a
substitute for a wide range of inpatient management, including both
acute and psychogeriatric care.

The overall rating for the scheme by professions taking part was
'successful' or 'very successful'. Similarly, patients and their relatives
reported the scheme 'helpful' or 'very helpful'. (It should be noted that
only 25 patients were able to be interviewed.) The daily cost of the
scheme was £43.05. This was higher than for a GP hospital (£30.32) and
partly acute hospital (£26.93), but less than for a mainly acute hospital
(£49.12).

The average length of stay for patients admitted to the study group
was 16.7 days. This was longer than for the average patient admitted to a
Peterborough partly acute hospital (9.7 days), but less than for GP
hospital patients (31.8 days) and mainly acute hospital patients. Consequently, the average cost per case in the hospital at-home scheme was

Mowat and Morgan (1982)

£721.61, which was less than for GP hospitals (£962.97), but more than for partly/mainly acute hospitals; £602.88 and £475.95 respectively.

5. General comments

By nature of its design, it is not possible to draw any firm conclusions from this study. The authors themselves acknowledged that comments presented should be regarded as 'tentative and impressionistic' and that 'particular caution should be exercised in drawing conclusions regarding relative costs'. No information was presented to validate the comparison of costs of hospital at home treatment with that incurred on behalf of 'average' patients in the acute and GP hospital sectors.

The study design would have been vastly improved by random allocation of patients into different types of inpatient management and comparing individual patient costs, rather than averages.

It would be expected that the hospital at home alternative would result in patients and relatives incurring higher costs than if care had been provided in a hospital and it is therefore unfortunate that this study only considered costs to the NHS. Although the possibility of any higher costs to patients and relatives needs to be set against the preferences for care at home.

Whilst the alternatives considered were only compared in revenue terms, a transfer from hospital to hospital at home management could result in capital savings for the NHS. The size of such savings would depend upon the scale of transfer of patients to hospital at home management and upon the extent of spare capacity within the NHS. (See Sections 3.2 and 7.1.2 of the companion volume.)

The average measure used in the study was also very limited. By nature of the study design, it is not possible to estimate whether the alternatives considered are equally effective. Even for the hospital at home group alone, there is no indication of the medical effectiveness of the treatments provided (except for the fact that patients did not have to be admitted to hospital).

8 Prescott, R. J., Cuthbertson, C., Fenwick, N., Garraway, W. M., and Ruckley, C. V. (1978). Economic aspects of day care after operations for hernia or varicose veins. *Journal of Epidemiology and Community Health* **32**, 222–5.

1. Study design

1.1. *Study question*

What are the costs or savings to the health service of day care surgery?(c)

1.2. *Alternatives appraised*

Day care surgery *versus* 48 hours after care in a surgical ward *versus* 48 hours after care in a convalescent hospital.

1.3. *Comments*

The authors noted that whilst their study compares day care with a two-day hospital stay, the mean duration of stay in 1975 was eight days for hernia repair and six days for varicose vein surgery. They argue that taking account of a longer length of stay would increase the potential savings to £100 per case.

2. Assessment of costs and benefits

2.1. *Enumeration*

The effectiveness of the alternative regimes had been reported in previous studies. The costs considered were from changes in health service use only; hospital costs after surgery, district nursing costs, GP costs, and ambulance service costs. The implications for patients and their families were referred to in the discussion but not explicitly evaluated.

2.2. *Measurement*

For the hospital care groups, information on individual resource use, such as drugs and tests, was recorded for each patient. Nursing and medical staff costs were allocated according to patient dependency levels. Ward and hospital overheads were allocated on a patient day basis. There was no attempt to identify which of these joint costs, if any, would be affected by a change to day case surgery (see Section 4.2.2. of the companion volume).

Use of health service resources after discharge was recorded for 21 days after the operation for both the day care and hospital care groups.

Prescott *et al.* (1978)

Data were collected by health service personnel and included contact time and travel time.

2.3. *Explicit valuation*

All cost estimates were based on 1975–76 wage and price costs to the NHS.

3. Allowance for differential timing and uncertainty

Differential timing was not relevant because of the relatively short time period and because no extra capital costs for providing the different facilities were incurred.

4. Results and conclusions

The total average costs of after care were reported to be £16 for patients discharged home, £38 for patients spending 48 hours in a convalescent ward, and £46 for patients cared for in surgical wards. The authors concluded that, in the particular situation of this study, 'there is not likely to be any immediate reduction in hospital expenditure because of the move towards early home discharge, but the turnover of patients will tend to increase, causing a reduction in the overall cost per patient, and a reduction in waiting lists'. (Of course, if this happens hospital expenditure is likely to increase.)

5. General comments

Only in the discussion did the authors begin to address the question of marginal, as opposed to average, costs. Problems may arise from shortages in community or convalescent care and the savings from reduced inpatient stay will be much less if surgical beds are under-utilized. The approach adopted in this study should be contrasted with Russell *et al.* (1977). Day case surgery for hernias and haemorrhoids: a clinical, social and economic evaluation, *Lancet* **1**, 844–7, summarized in the first volume of *Studies*, 1981. (See also the detailed comments on elective surgery in the companion volume, Section 7.1.)

9 Stilwell, J. A. (1979). Relative costs of home and hospital confinement. *British Medical Journal* **2**, 257–9.

1. Study design

1.1. *Study question*

What are the comparative costs of confinements at home or in hospital for normal births?(a)

1.2. *Alternatives appraised*

Booked confinement at home *versus* GP hospital with 48-hour stay *versus* consultant obstetric unit with 48-hour stay.

1.3. *Comments*

There may be an implicit assumption that outcomes are equivalent for each location, making this a cost-effectiveness study. This is still a contentious issue, even for normal births, and therefore the study is treated as a costing exercise.

2. Assessment of costs and benefits

2.1. *Enumeration*

The costs considered were those from the use of health service resources, plus other public sector costs and costs to the family. Health service costs were taken to include antenatal and postnatal care, not just delivery. Where working days were lost, the net wage was treated as a family cost and the lost tax yield as a public sector cost. Changes in the use of non-marketed time were not included.

2.2. *Measurement*

Data on the use of health service resources were obtained from midwives, family doctors, and hospital notes. The cost of a 48-hour stay in a GP unit had to be estimated from average costs for a four day stay. Estimation of the costs for the consultant unit required an apportionment of costs between normal and abnormal cases. Family costs were derived from a questionnaire giving information on days lost from work, hospital visits, etc., but not actual expenditure.

2.3. *Explicit valuation*

Market values were the main source of cost estimates. Wage rates were adjusted to take account of the average unemployment level.

Stilwell (1979)

3. Allowance for differential timing and uncertainty

Differential timing was not relevant in this context of the study.

4. Results and conclusions

The cost differences were found to be small. The average cost per birth was £257.36 for home confinements, £270.67 for GP hospital deliveries, and £330.17 for the consultant unit. In addition, community midwife visits to the home delivery group and midwife attendance at delivery compensated for avoided hospital costs. Time lost from work was the most significant family cost and this was highest for the consultant unit group.

5. General comments

As there are relatively few home deliveries, the numbers studied were low (22 home deliveries with a matched sample of GP deliveries). Therefore the results may not be representative. Although the different pattern of antenatal care for the home group has an effect on the cost differences, there was no discussion of why (or whether) this was necessary, i.e. it is not clear whether each form of care was being delivered as efficiently as possible.

The author pointed out that the marginal costs of each type of care were likely to be lower than the average costs presented in the study. It is the marginal costs that would be relevant in considering any changes in the balance of care.

O Weisbrod, B. A., Test, M. A., and Stein, L. I. (1980). Alternative to mental hospital treatment. II. Economic benefit–cost analysis. *Archives of General Psychiatry* **37**, 400–5.

1. Study design

1.1. *Study question*

What are the costs and benefits of conventional hospital-based treatment for mentally ill patients compared to an experimental community-based programme?(c)

1.2. *Alternatives appraised*

Traditional hospital-based treatment (institutional care plus outpatient visits) *versus* community-based care ('training in community living').

1.3. *Comments*

The economic appraisal was linked to a randomized controlled clinical trial. Details of this, plus the treatment alternatives, are given in another article in the same journal. (Stein, L. I. and Test, M. A. (1980). *Archives of General Psychiatry* **37**, 392–7.)

2. Assessment of costs and benefits

2.1. *Enumeration*

Costs for hospital and community patients included direct and indirect treatment costs, law enforcement costs, living expenses and family burden costs. Earnings from competitive employment and sheltered workshops were taken as the measure of benefits for which monetary estimates were made, although changes in clinical symptomology were also considered. The results were expressed as average costs or benefits per patient for each group.

2.2. *Measurement*

Data were collected over 12 months. Hospital treatment costs took account of operating costs, capital costs and a percentage rate of return on the market value of the part of the hospital used for patients. Outpatient costs were calculated by reference to staff time spent with patients. Community costs included all the costs associated with the experimental centre and the value of administrative services given without charge. The cost per patient was calculated assuming a near-

Weisbrod *et al.* (1980)

maximum patient population. (At the early stages of the study the centre was operating well below full capacity.)

Indirect treatment and law enforcement costs for both groups were obtained from data collected by interview from patients and their families, and from information provided by the appropriate agencies. Living expenses for both groups were derived from agencies or patients.

To measure benefits, data on earnings and sheltered workshop income were obtained from patients and verified by the employers.

2.3. *Explicit valuation*
Market values and normal fees and charges were used to estimate costs.

3. Allowance for differential timing and uncertainty

Hospital treatment costs included capital costs (to which depreciation rates based on market values were applied) and a 9 per cent rate of return on the value of that part of the hospital used by the control group of patients was assumed. No sensitivity analysis was performed.

4. Results and conclusions

Direct treatment costs under the experimental programme were higher than for the control group, but all other costs, i.e. indirect treatment, law enforcement, maintenance, and family burden costs were lower. Overall, therefore, the total cost per patient was US$797 (11 per cent) higher for patients in the experimental programme, but the benefits accruing to patients in this group were also higher (US$1196). Taking valued benefits minus valued costs, the experimental group imposed a lower cost on society than the hospital control group. (The experimental group also had better clinical symptomology and higher patient satisfaction after the 12 months.)

5. General comments

The favourable outcome for the experimental group relied on earnings for the experimental group being higher than those for the control group, and outweighing the difference in costs between the two groups. In times of high unemployment, for example, these benefits may not accrue and also the difference in overall reduced cost for experimental over control group was very small. A range of values for the benefits would have given levels at which the control group cost would have been lower than for the experimental. On the other hand, many benefits were not given a monetary value and they appeared to favour the experimental group. Such variations could therefore affect the stated results and hence more sensitivity analysis, and associated policy implications should have been pursued.

81 Weissert, W. G., Wan, T. T. H., Livieratos, B. B., and Pellegrino, J. (1980). Cost-effectiveness of homemaker services for the chronically ill. *Inquiry* **17**, 230–43.

1. Study design

1.1. *Study question*

Would homemaker services reduce levels of institutionalization?

Would homemaker services increase or maintain physical functioning at levels as high or higher than existing care options?

What would be the effects of such services on the use of other Medicare services, and the cost implications?(c)

1.2. *Alternatives appraised*

Receipt of homemaker services plus other care options *versus* other care options alone. (These included hospital and skilled nursing inpatient care, home health visits, physician visits, and a variety of other ambulatory services.)

1.3. *Comments*

Patients were carefully selected from four cities, and were randomized into two groups and followed-up over one year. Five assessments were made and the resulting complex clinical trial was also analysed with respect to cost.

2. Assessment of costs and benefits

2.1. *Enumeration*

The study considered costs to the health service. Benefits were measured as changes in health status and reductions in demands on health services. Statistical analysis using regression techniques was performed to determine the effect of various prognostic and service utilization variables on outcomes.

2.2. *Measurement*

Clinical benefits such as changes in morbidity, mortality, and levels of quality of life were measured in the assessments mentioned above.

The costs of the homemaker services, hospital care, nursing home care, and out-of-pocket expenses were calculated for both patient groups, and the yearly costs of the homemaker and Medicare services were derived.

All costs were expressed as average costs only.

Since all patients in the study were defined as chronically ill, productive output was not taken into account.

2.3. *Explicit valuation*

The cost data were derived from four sources:

- (a) Medicare billing files were consulted.
- (b) Providers of homemaker services calculated their own costs, monitored by the Research Unit implementing the study.
- (c) Patients were asked to calculate their out-of-pocket expenses and private insurance costs.
- (d) Medicaid reimbursement records were consulted.

  Costs derived under (c) were only to be used if statistically different between the groups, and in fact were not used. Very few patients were covered under (d) and therefore these costs were not considered.

3. Allowance for differential timing and uncertainty

Differential timing was not relevant in the context of this study. Regression analysis on the clinical data estimated the effect of changing particular variables, and analysis including or excluding missing and contaminated cases were performed.

There was no sensitivity analysis on the cost estimates.

4. Results and conclusions

- (a) The most striking finding of the multi-stage analysis was that the use of homemaker services significantly affected the mortality rate, but not the physical functioning or rate of institutionalization of patients.
- (b) The costs of providing homemaker services averaged US$7.61 per hour, or US$2290 per annum. When added to the Medicare costs, the costs for homemaker patients were on average US$3432 more than the costs of the control group. Homemaker services appeared to increase the costs of existing services rather than serving as a substitute for them. However, some of the higher costs were due to the approximately 57 patients apparently kept alive by homemaker services.

The authors concluded that these results 'do not prove that these services constitute a cost-effective alternative in long-term care'.

Weissert *et al.* (1980)

5. General comments

The authors considered that there was a need to investigate further the findings that homemaker services reduce mortality rates. It would be necessary to consider the duration and quality of these additional life years.

32 Wright, K. G., Cairns, J. A., and Snell, M. C. (1981). *Costing care*. Social Services Monographs: Research in Practice. University of Sheffield Joint Unit for Social Services Research, Sheffield, UK.

1. Study design

1.1. *Study question*

What are the relative costs of care in different settings for elderly people with similar requirements for help (or dependency)?(a)

1.2. *Alternatives appraised*

Care in long stay hospitals wards *versus* local authority residential care *versus* own homes in the community. (See 1.3 below.)

1.3. *Comments*

The basic objective was to estimate the relationship between cost of care and level of dependency in the alternative settings, with a view to providing part of the information required to make judgements on the appropriate balance or mix of care for the elderly population. (See the companion volume, Section 7.4, for more discussion of this approach.)

2. Assessment of costs and benefits

2.1. *Enumeration*

The costs considered included those falling on the health service and local authorities, plus those falling on the elderly and their families (in personal living expenses and in providing informal help). Both capital and recurrent costs were considered.

The dependency items considered related mainly to physical dependency, i.e. activities of daily living.

2.2. *Measurement*

Special emphasis was placed on the measurement of dependency by the Guttman scaling method. In the community care setting the range and frequency of services received by the elderly were recorded; also estimates were obtained from 'principal helpers' of their input. In the residential care alternative, estimates were obtained from carers of the time input required by dependency level. In the hospital setting average long stay hospital costs were used.

Wright *et al.* (1981)

Measurements were made in three geographical areas to examine the impact of urbanization, population density, and the prices of commodities and labour.

2.3. *Explicit valuation*

The authors point out the fact that many of the prices observed for services may not reflect true opportunity costs. Also, much of the capital expenditure (say, in hospitals) had already been 'written off'. However, in general a pragmatic approach was taken to costing health service and local authority resources, frequently relying on the prices observed. Details are given in an appendix.

3. Allowance for differential timing and uncertainty

Capital expenditures were annuitized by using a discount rate of 7 per cent, the recommended UK Treasury rate at the time of the study.

No formal sensitivity analysis was performed, but at a number of points in the study the impact on estimates of alternative assumptions is discussed.

In addition the results of the study were compared with those obtained by other researchers.

4. Results and conclusions

For the low dependency groups, it was found that many elderly persons appeared to be placed in residential homes, even though this was at a higher cost than community care for the equivalent dependency level. The authors attribute this to the importance placed on factors such as personal risk, household tension, and prognosis by those care givers who typically decide on placement of the elderly. (That is, one cannot say from these data that the elderly are being inefficiently placed, since we have no data on relative effectiveness of the care modes.)

In the higher dependency groups, the main conclusions were that ways should be considered to expand the provision of the less costly forms (residential care and community care). In particular it was suggested that help (financial or otherwise) could be given to those families supporting elderly at home, as this would save the community resources.

5. General comments

The study contained a good general discussion of the measurement problems (both in estimating dependency and cost) faced by researchers in this field.

Also the study contained a thorough discussion of costs by various viewpoints, especially the contrast between the government ('public expenditure') and the societal ('economic') perspectives.

SECTION 6: Alternatives in health service organization

Introduction

This section brings together studies that are concerned with aspects of health service organization. A growing number of studies deal not with choices in medical technology but with choices in methods of organizing and delivering clinical and non-clinical services. Eighteen studies are reviewed and the range of issues covered is broad, reflecting something of the scope that exists for evaluation. Although most of the studies discuss different questions, they can be placed into one of the three following categories:

    (i) the management of clinical or support services
    (ii) the use of supplies
    (iii) the substitution of different providers of health services.

The last category is the smallest, containing only three studies. These are concerned with the potential that exists in some circumstances for substituting either self care or nursing care for primary care by physicians (Zapka and Averill 1979; Greenfield *et al.* 1978) and for substituting parent care for nursing care of children (Evans and Robinson 1983). From the economists' perspective, the substitution between different types of labour in different settings is a potentially rich field of study. Whilst there have been studies based on aggregate statistical analysis (for example Reinhardt 1975), the fact that there are so few practical examples of evaluation of specific substitution possibilities suggests that there may be difficulties in obtaining co-operation to set up such studies. One study of the introduction of nurse practitioners in Canada has shown that the incentives for producers can be perverse. The family practices that participated in the experimental programme suffered a financial loss as the nurse practitioners could not generate enough extra revenue (in health insurance plan reimbursements) to cover their salary (Spitzer 1978).

The studies of supplies services provide examples of three types of questions that can arise. Alternative types of clinical materials are evaluated in four studies (Adar and Pliskin 1980; Culyer *et al.* 1983; Fineberg *et al.* 1980; Goode *et al.* 1979). The examples are vascular grafts, wound dressings, and anaesthetic gases. The potential for using or re-using disposables arises frequently and studies of syringes and haemo-dialysis equipment are reviewed (Strathclyde Diabetic Group 1983; Banester *et al.* 1982). The final type of example looks at different methods of supplying oxygen in patients' own homes (Lowson *et al.* 1981).

There are eight studies on the management of services covering diverse topics. In the acute sector, there are examples of studies examining the regionalization of heart surgery (Finkler 1981), alternative ways of providing emergency care for cardiac arrest (Hallstrom *et al.* 1981), second opinion consultations before elective surgery (Ruchlin *et al.* 1982), and different 'return to work' regimens after hernia repair (Bourke *et al.* 1982). Two studies are concerned with organizational initiatives in community care; emergency alarms (Ruchlin and Morris 1981) and co-ordinating centres (Gunning-Schepers *et al.* 1984). Finally, there are studies evaluating the use of laboratory tests (Stilwell *et al.* 1980) and the organization of contact lens provision (Woodward and Drummond 1984).

Particular methodological problems in this area

The main methodological issue to arise in this area is the influence that social, economic, demographic or organizational features of particular settings may have on the costs and outcomes of alternative methods of delivering or organizing health care. Finkler (1981) provides a good example of how to approach this problem. In evaluating the regionaliza-tion of heart surgery he draws attention to three factors that will influence the results of the study; population density, the method of payment of physicians, and whether or not a dedicated theatre is provided. Lowson *et al.* (1981) also examined different conditions for the supply of domiciliary oxygen. However, most of the studies reviewed are concerned with a particular problem and do not address the issue of whether the results can be generalized to other settings.

Although they also arise in other areas, some other general points are worth noting. The organization of services often has effects on patient costs and these are frequently ignored or handled badly. For example, the re-use of diposables involves the patient's time in sterilization which was not taken into account in the examples here (Banester *et al.* 1982; Strathclyde Diabetic Group 1983). However, the potential for saving the patient's time was one of the main motivations for the study of contact lens provision (Woodward and Drummond 1984). In other instances, the costing methodology was poor throughout (Bourke *et al.* 1981; Goode *et al.* 1979). Where the alternative methods of organization have effects on inpatient stay, these have invariably been costed in average rather than marginal terms (Goode *et al.* 1979; Gunning–Schepers *et al.* 1984; Ruchlin and Morris 1981). Finally, there was relatively little use of sensitivity analysis in circumstances where it would have been appropriate. There are two good examples of the proper use of sensitivity analysis in the section; the more familiar use of a range of values to overcome missing or imperfect data (Culyer *et al.* 1983) and the more recent practice of testing results for statistical significance (Greenfield *et al.* 1978).

Current state of the art

There is an inherent danger that studies of organizational issues will be carried out from the viewpoint of the particular decision maker, rather than society as a whole. There are certainly examples of this phenomenon in the section (Goode *et al.* 1979; Greenfield *et al.* 1978). Changing the organization of a service is often motivated by the desire to make better use of the resources available to that service and lack of concern for the impact on other services or the community at large is not, therefore, surprising. Some of the studies do avoid this problem, particularly if the study has been carried out by disinterested parties.

The problem of 'tunnel vision' with respect to costs and benefits tends to be greatest when appraisals are carried out by those involved in providing a service. There also seems to be a related tendency, in some cases, to set out to prove a particular point of view. This is exemplified by studies which only examine one option in depth (for example Goode *et al.* 1979). In adopting a narrow approach to the problem, the incentives for other parties to co-operate or not co-operate tend to be ignored, even when it is explicitly recognized that this will affect the implementation of change (Banester *et al.* 1982). In this area, perhaps more than any other, the costs of managing change or inducing co-operation might be explicitly considered and set against potential benefits.

In considering gaps in the current literature, it is clear that many more studies could be carried out in this area. However, the published literature is not necessarily a good guide to the activity taking place. It is certainly true that, in the UK, more work of this type is carried out, particularly in non-clinical areas, such as catering, energy conservation, and transport organization (Ludbrook 1984). The dissemination of the results of such studies is poor, mainly because the work is carried out by health service staff who see little incentive to publicize their work more widely.

Although the inclusion of unpublished work improves the situation, the amount of economic evaluation of organizational issues is still not great. It is still the case that in the UK most decisions, regarding for example, the automation of laboratory services, the microfilming of hospital records or the reorganization of laundry services, are taken without any explicit economic evaluation. The main area where economic appraisal could make an impact, however, is in service planning. Recent initiatives in the planning requirements for NHS capital schemes should mean that there will be greater use of economic appraisal in this area. In other countries, and in other systems, there is little evidence to suggest that economic appraisal, as opposed to financial appraisal, is being introduced into health service planning.

Contribution to decision making

In practice, the best test of the contribution made to decision making is whether better decisions are taken and the efficiency of health service delivery improved. This can be difficult to detect and the time lapse between an appraisal being carried out and the final decision being taken can be considerable.

A good example of this is the case of oxygen concentrators in the UK (Lowson *et al.* 1981). A decision to use oxygen concentrators more widely has only just been taken following protracted negotiations at national level. Two factors complicated the implementation of the more cost-effective delivery system.

First, at the time when the study was carried out, the supply of oxygen concentrators would have to be funded from a different budget than the supply of oxygen cylinders and cash could not be transferred between the two budgets. Secondly, the NHS had a monopoly supplier of oxygen. It was not clear, therefore, whether the saving to the NHS would be realized from switching to oxygen concentrators if the supplier of other oxygen products changed its pricing structure. Similar sorts of issues, relating to the distribution of additional costs or benefits, arise in many studies and clearly affect the implementation of study results.

It is rare for any published follow-up of an economic appraisal to appear. One recent exception was a report on contact lens provision (Astin 1984) following an earlier economic appraisal (Woodward and Drummond 1984). The actual service situation was compared with the trial conditions and assumptions of the study. The changes which had been implemented appeared to have worked well but some minor revisions of the system were suggested as a result of the practical experience.

Appraisal studies are still useful, even if their advice appears to be ignored. Generally, appraisals make more information available to decision makers and highlight the costs of pursuing other courses of action (or inaction). However, most of the studies in this section clearly do aim to affect decision making, at least locally. Many of them were carried out by, or in conjunction with, local providers of services. Whilst this has advantages, both in terms of access to information and in getting decisions implemented, there is a danger that studies which have no external input may display a degree of bias that renders them useless for decision making purposes. This is a particular problem in this area because the results may genuinely be dependent upon local circumstances. The expert advice which may be needed, and which can be compared to clinical evidence needed in other types of study, may therefore be less open to peer review and independent criticism.

Other problems which were highlighted earlier may reduce the value of study results in a wider decision making context. The organizational setting of a study may imply that the results cannot be achieved in other settings, although the way of structuring the problem may still be useful. The study results might have more impact if this problem were addressed explicitly by the study authors. Finally, the issue of incentives for change is almost universally ignored, but decision makers have to be concerned with the problem of implementing proposals.

References

Astin, C. (1984). Aphakia contact lens fitting in a hospital department, *Journal of the British Contact Lens Association* 7,3, 164–8.

Ludbrook, A. (1984). *Economic appraisal in the NHS: a survey.* Health Economics Research Unit Discussion Paper 01/84. University of Aberdeen.

Reinhardt, U. E. (1975). *Physician productivity and the demand for health manpower.* Ballinger, Cambridge Mass.

Spitzer, W. O. (1978). Evidence that justifies the introduction of new health professionals. In *The Professions and Public Policy* (P. Slayton and M. J. Trebilcock, eds). University of Toronto Press.

3 Adar, R. and Pliskin, N. (1980). Cost analysis of the utilization of new vascular grafts. *Metamedicine* **1**, 213–23.

1. Study design

1.1. Study question

What are the costs associated with the utilization of different types of material for vascular grafts?(c)

1.2. Alternatives appraised

The dacron graft *versus* expanded microporous polytetrafluoroethylene (PTFE) *versus* the modified human umbilical cord vein (UV).

1.3. Comments

The study employed a decision tree to examine the problem of a hypothetical 60-year-old male patient undergoing a vascular graft, who could be followed-up for a further five years. Probabilities for the outcomes at the end of each year were assessed and all alternative paths were costed.

2. Assessment of costs and benefits

2.1. Enumeration

The costs considered were those of the graft itself (the actual cost of surgery was the same in each case), the cost of amputation, and the cost of disability. Each of these was weighted by the probability of outcome at each stage. The cost of each path could be summed to a total cost, by multiplying up the cost and probability of each stage. Details of the items considered in each cost category were not given.

2.2. Measurement

The probabilities were derived from personal knowledge and a literature survey. The authors did not say how the costs were arrived at. No specific mention was made of lost wages, nor of the costs of keeping the disabled person in hospital or home. The costs of death, and the capital costs were not discussed. There was no discussion of average and marginal costs. To simplify the calculation, the decision tree was 'pruned', and the costs of disability were averaged over two years.

2.3. Explicit valuation

Market values appear to have been used for costs, though it was unclear how the cost of disability was assessed.

3. Allowance for differential timing and uncertainty

A discount rate of 10 per cent was applied, but the effect of alternative discount rates was not assessed. Although the authors acknowledged that the probabilities of certain actions taking place might be inaccurate and therefore could be queried, no sensitivity analysis was attempted.

There was no discussion of the probabilities associated with different patient populations.

4. Results and conclusions

Although the newer graft was more expensive to perform, the total cost over five years of this graft was only 31 200 Israeli lira, whilst the total cost of the dacron graft (a cheaper operation to perform) was 69 600 Israeli lira. In other words, the actual cost of the graft was a small part of the total expenditure over five years on these patients.

5. General comments

The methodology adopted here (that of a decision tree applied to a theoretical case) could be utilized in other situations in which new treatments appear more costly than the established methods. The study highlights the importance of evaluation before procedures are accepted or rejected, and picks out areas in which more data should be collected, e.g. probabilities, from say, clinical trials. More information is also needed on the benefit side.

84 Banester, H. K., Driscoll, R. D., Greenwood, R. N., and Cattell, W. R. (1982). Reuse of haemodialysis equipment: convenience and cost effectiveness, *British Medical Journal* **285**, 473–4.

1. Study design

1.1. Study question

Is the re-use of disposable dialysis equipment cost-effective?(c)

1.2. Alternatives appraised

Single use *versus* multiple use (maximum six times).

1.3. Comments

In addition, the study took account of the different types of equipment which patients were using; Kiil (non-disposable) dialysers and disposable dialysers with either manual or automated re-use.

2. Assessment of costs and benefits

2.1. Enumeration

The re-use of dialysis equipment involves consideration of the safety and convenience for the patients, as well as the actual costs of materials involved in either single use or re-use. The study considered both the re-use of disposable dialysers and the re-use of arterial and venous blood lines. Different capital costs for machines were not included.

2.2. Measurement

The basis of the study was a survey of re-use practice amongst the 72 patients on home dialysis at one renal unit. On the safety aspects, the authors refer to the results of other studies. Although convenience to patients was identified as an issue, no attempt was made to measure the amount of patients' time involved in re-use of equipment.

2.3. Explicit valuation

Materials were valued by their cost to the NHS.

3. Allowance for differential timing and uncertainty

Differential timing was not relevant as capital costs had been excluded. No sensitivity analysis was used and the distribution of dialyser re-use around the mean was not reported.

Banester *et al.* (1982)

4. Results and conclusions

Single use of dialysis equipment incurred a materials cost of £1872 per patient per year. Using all disposable equipment a maximum of six times reduced the cost to £637, including the cost of materials required for re-use.

Of the patients with disposable equipment, those with an automated re-use facility re-used their equipment more times on average; 4.9 *versus* 4.6 for dialysers, 5.0 *versus* 2.7 for arterial and venous lines. This at least suggested that patients' time was a factor in determining re-use practice and should not, therefore, have been ignored. The authors claimed that the capital cost of equipment for automated re-use would be covered by savings within the first year but did not report the figures that substantiate this.

5. General comments

85 Bourke, J. B., Lear, P. A., and Taylor, M. (1981). Effect of early return to work after elective repair of inguinal hernia: clinical and financial consequences at one and three years. *Lancet* **2**, 623–5.

1. Study design

1.1. *Study question*
What are the clinical and financial effects of early return to work after elective repair of inguinal hernia?(c)

1.2. *Alternatives appraised*
Early return to full working capacity *versus* return to full working capacity as supervised and determined by patients' GPs.

1.3. *Comments*
Patients were randomly allocated between the two groups. Within the groups, patients were classified into four sub-groups according to occupation. Patients were assessed after one year and three years.

2. Assessment of costs and benefits

2.1. *Enumeration*
The clinical effect of early return to work was defined in terms of the recurrence rate assessed at one and three years. Financial consequences were limited to gains and losses for the individuals whilst off work. Although health service costs in the first two to three weeks would be the same for both groups, there may be differences in follow-up costs, but these were not considered.

2.2. *Measurement*
Patients were randomly allocated between the early return group and GP supervised return group two to three weeks after operation. They were independently assessed for recurrence after one year and three years. Information about financial gains and losses was recorded at the one year review, which may imply some degree of recall error. Few details were given but the financial consequences seem to have been measured purely from the patient's perspective, after transfer payments had been taken into account. Actual loss of production was not measured, or even proxied by loss of earning, therefore.

2.3. *Explicit valuation*
Monetary values of gains and losses whilst off work were used.

Bourke *et al.* (1981)

3. Allowance for differential timing and uncertainty

A three year follow-up was used to check the results of the one-year follow-up. Results were calculated separately for different categories of workers; retired, light work, intermediate work, heavy work, and self-employed.

4. Results and conclusions

The recurrence of hernias was not significantly different for the two groups. There was a difference in time off work between the control and experimental groups for all groups of workers, but not for the retired. The financial differences between the two groups were not explicitly calculated. The median income loss per week was stated, and the study implied that therefore, if patients returned to work earlier, the income loss would be lower, particularly for those who were self-employed.

5. General comments

The clinical evidence was well documented. The financial evidence was almost non-existent and some types of resource use were ignored. Therefore, although the study may suggest that early return to work is advantageous the results must be viewed as tentative in terms of the economic outcomes.

36 Culyer, A. J., MacFie, J., and Wagstaff, A. (1983). Cost-effectiveness of foam elastomer and gauze dressings in the management of open perineal wounds. *Social Science and Medicine* **17**,15, 1047–53.

1. Study design

1.1. *Study question*

What is the relative cost-effectiveness of foam elastomer dressing for open perineal wounds?(c)

1.2. *Alternatives appraised*

Foam elastomer dressing *versus* conventional gauze dressings.

1.3. *Comments*

The authors commented on the difficulties that resulted from the failure to incorporate the economics component into the initial design of the clinical effectiveness study. The decision to include economic analysis was taken after the trial had been carried out and although the data problems were overcome in this case, it is stressed that good research design requires early collaboration.

2. Assessment of costs and benefits

2.1. *Enumeration*

The costs for each type of dressing included the costs of materials, other inpatient costs, non-material costs after discharge, and the costs of complications. The cost of patient time for outpatient visits was not included but transport costs included the driver's time, regardless of who it was.

2.2. *Measurement*

Effectiveness was based on a prospective randomized controlled trial, previously carried out on a group of 50 patients 14 days after surgery. Most of the data on resource use also came from this trial and were supplemented with estimated figures where data were not available. Costs which were known to be the same for both groups were ignored, e.g. hospital 'hotel' costs. The paper provided very detailed information on the costing of the two treatments.

263

Culyer *et al.* (1983)

2.3. *Explicit valuation*

Costs were valued according to NHS wage and price costs or market prices. Drivers' time was valued by national average earnings.

3. Allowance for differential timing and uncertainty

Discounting was not required.

Three estimates were presented—high, medium and low—based on one standard deviation above and below the mean per patient where data were available and a spread of estimates where data were not available. The authors recognized that in some cases the missing data were replaced by 'informed guesses' and attempted to ensure the robustness of the results despite these difficulties. Throughout the study the costing procedure was consciously biased against the foam elastomer treatment.

4. Results and conclusions

There was no significant difference between the two treatments in time taken for wound healing or in length of inpatient stay. However, patients treated with foam elastomer dressings required less pain killing drugs.

The cost estimates per patient for foam elastomer dressings were £70.50 (low), £162.10 (medium) and £422.00 (high). The corresponding estimates for gauze dressings were £146.10, £417.60 and £984.40. Given that the study had tended to bias the costs against the foam elastomer dressings, the authors concluded that the relative cheapness of these dressings was a fairly robust result.

5. General comments

This study is a particularly good example of how to handle a situation in which the data are less than ideal; a problem encountered all too frequently. On the specific issue of wound dressings this study can be contrasted with Goode *et al.* (1979) summarized in this section.

Evans, R. G. and Robinson, G. C. (1983). An economic study of cost savings on a care-by-parent ward. *Medical Care* **21**(8), 768–82.

1. Study design

1.1. *Study question*

What are the cost savings arising from care-by-parent for children in hospital?(c)

1.2. *Alternatives appraised*

Care-by-parent (CBP) *versus* traditional nursing unit (NU) care.

1.3. *Comments*

"This article describes a study at Children's Hospital in Vancouver in which actual resources used and costs are compared for similar episodes of care provided in the CBP ward and in inpatient wards."

2. Assessment of costs and benefits

2.1. *Enumeration*

The effectiveness of CBP was taken as given. The authors reported that the medical and psychologic advantages to patient and family were well established. This study concentrated on the costs. All types of cost, incurred in the hospital were included in the comparison. Parent costs were not included. The authors argued that as the parents had the choice of a CBP admission or a NU admission, their choice of the former indicated that the benefits (to the parents) were greater than any costs. Therefore, the advantages of CBP were in fact undervalued.

2.2. *Measurement*

The study was based on a retrospective cost comparison between a group of children admitted to the NU but who would have been suitable for admission to the CBP ward and patients with the same diagnoses treated in the CBP ward. Actual resource use was extracted from records of hospital charges and used as the basis for allocating costs.

2.3. *Explicit valuation*

Costs for each type of care were arrived at by allocating total hospital costs and therefore must be based on market prices faced by the hospital. References were cited for details of the costing procedure.

Evans and Robinson (1983)

3. Allowance for differential timing and uncertainty

Capital costs were converted to annual equivalents. The discount rate used was not given.

Although no explicit sensitivity analysis was performed, the authors did identify the savings attributable to different levels of resource use and to different unit costs. This assisted in interpreting the results.

4. Results and conclusions

The CBP ward resulted in savings ranging from Canadian $45 per case for tonsils and adenoids to $670 for endocrine cases. Only neurology patients showed no saving (in fact a slight increase). The savings resulted from, in order of importance, shorter stays, reduced ward costs, and reduced use of diagnostic tests. Given the patient matching, the authors argued that early discharge was an effect of the CBP ward, possibly due to greater physician confidence in the parents.

The use made of the CBP ward in 1975 yielded savings of over $46 000. Caring for all suitable patients in the CBP ward would have yielded additional savings of $194 000 provided that the CBP ward substituted for NU care and did not lead to the generation of additional patients.

5. General comments

The authors discuss various problems including whether savings are actually realized (see Evans and Robinson, 1980, summarized in Section 5), comparability of controls and the adjustment of hospital costs to the types of changes induced, given that some cost elements are fixed in the short run.

88 Fineberg, H. V., Pearlman, L. A., and Gabel, R. A. (1980). The case for abandonment of explosive anesthetic agents. *New England Journal of Medicine* **303**, 613–17.

1. Study design

1.1. *Study question*
Should the use of explosive anaesthetic agents be discontinued?(d)

1.2. *Alternatives appraised*
Continued optional use of explosive anaesthetics, such as cyclopropane, *versus* a complete ban.

1.3. *Comments*
The use of explosive anaesthetic agents was shown to be declining but an explicit comparison between banning their use now, and declining levels of use in the future was not carried out.

2. Assessment of costs and benefits

2.1. *Enumeration*
'The potential benefits from continued use of explosive anesthetic agents include their value in teaching, professional freedom, familiarity with inflammable agents, pharmacologic advantages in selected patients and the desire to retain an option to deal with possible problems in the future.'

The arguments against retaining their use were lack of general pharmacological advantage, loss of familiarity with their use, risk of explosion, and cost. The additional costs of using explosive anaesthetics came from safety measures to reduce static electricity and make electric equipment and outlets explosion proof.

2.2. *Measurement*
The costs of safety measures were estimated from the recommendations of the National Fire Protection Association, published data and discussions with various authorities. Sunk costs were rightly ignored in the calculations. The other costs and benefits were simply discussed, with references to existing literature.

Fineberg *et al.* (1980)

2.3. *Explicit valuation*

Few details of the costing procedure were given. It must be assumed that market prices were used for valuation.

3. Allowance for differential timing and uncertainty

No discounting was carried out. The study had not considered, as it should, alternative flows of costs and benefits over time resulting from different policies.

4. Results and conclusions

The authors estimated that safety measures relating to the use of explosive anaesthetics cost US$9.4 million per year. They stressed that safety measures incur fixed costs for each hospital that provides for the use of these anaesthetics. As the frequency of use is falling, the cost per anaesthetic administered will continue to rise. The authors concluded that the weight of argument favours a ban on explosive anaesthetics.

5. General comments

The study was careful to enumerate all potential costs and benefits, even where these could not then be valued.

89 Finkler, S. A. (1981). Cost-effectiveness of regionalization—further results for heart surgery. *Health Services Research* **16**(3), 325–33.

1. Study design

1.1. Study question

Is the concentration of heart surgery in regional centres cost-effective?(c)

1.2. Alternatives appraised

Existing provisions *versus* centralization.

1.3. Comments

The study looks at potential economies of scale to be realized from centralization of provision in a suburban area with four hospitals. An earlier study was carried out in a large city (S. A. Finkler (1979). Cost-effectiveness of regionalization: the heart surgery example. *Inquiry* **16**(3), 264–70).

2. Assessment of costs and benefits

2.1. Enumeration

The costs considered were the additional hospital costs involved in providing facilities for heart surgery and the travel costs for patients and visitors, in terms of both time and money. Attention was centred on fixed hospital costs as this was where economies of scale would arise. In this study, physician costs were excluded because the method of payment, fee-for-service rather than salary, made them a variable cost to the hospital.

2.2. Measurement

The additional resources involved in providing facilities for heart surgery were identified for one hospital, using a method which the author refers to as component enumeration. This process derives marginal costs; joint costs and overheads were excluded as having no bearing on possible economies of scale. Although the methodology was discussed quite fully, few details were given of the costs for this study. Details of the measurement of travel costs were not given.

2.3. Explicit valuation

Sources of costs were not detailed in the paper. It must be assumed that wage and price costs to the hospital were used.

3. Allowance for differential timing and uncertainty

Discounting was not mentioned and it is not clear whether any was carried out. In a study of this nature, capital and equipment costs should be discounted to enable comparisons to be made.

The author discussed the effect that different hospital management practices can have on outcome. Fee-for-service remuneration of doctors rather than salaries made physician costs variable rather than fixed and reduced the scale at which the minimum average cost per patient was achieved. Similarly, the practice of dedicating operating theatres exclusively to heart surgery would increase the volume of operations indicated for regional centres to be viable.

4. Results and conclusions

'The principal result of this study is that strong economies of scale do exist in the production of heart surgery. The relatively flat portion of the average cost curve is reached at a volume of approximately 300 procedures per centre per year. (By 'flat portion' we meant that the additional reduction in cost per patient as volume increases is relatively small.)'

The potential savings for the four hospitals in the study area were not explicitly calculated. However, the author reports that on the basis of this study, 88 per cent of California hospitals are economically inefficient in the production of heart surgery. The earlier study, in an urban area, had found the minimum efficient scale to be 500 operations per centre per year. There is a clear need, therefore, to consider the precise setting before applying the results of this form of study to other problems and other locations.

5. General comments

Outcomes were not specifically considered in this study. However, the author does refer to literature concerning the advantages of regional centres in terms of the quality of care and outcome. (Luft, H., Bunker, J. and Enthoven, A. (1979). Should operations be regionalized? The empirical relation between surgical volume and mortality. *New England Journal of Medicine* **301**(25), 1364–9.)

0 Goode, A. W., Glazer, G., and Ellis, B. W. (1979). The cost-
effectiveness of Dextranomer and Eusol in the treatment of
infected surgical wounds. *British Journal of Clinical Practice* **33**,
325–8.

1. Study design

1.1. Study question

What is the most cost-effective way to treat infected surgical wounds?(c)

1.2. Alternatives appraised

Twice daily instillation of Debrisan granules *versus* twice daily dressings
with Eusol and paraffin-soaked dressings of ribbon gauze.

1.3. Comments

Despite the stated objectives, the study was more concerned with the
relative effectiveness of the two methods of treatment than with their cost
implications. The cost-effectiveness of the alternatives could not be
inferred directly from the information presented in the study.

2. Assessment of costs and benefits

2.1. Enumeration

The costs considered were the costs of the two treatment materials. Any
differences in staff time involved in administering these treatments were
not commented on. Changes in costs incurred outside hospital care were
not considered. The benefits were in terms of the earlier closure of the
wound and consequent early discharge of the patient.

2.2. Measurement

Data were collected for 20 patients, randomly allocated to treatments
after infection had occurred. The only measurements reported in the
paper concern the time to closure of the wound and the difference in
length of stay. The cost of saving resulting from early discharge is not
measured. The only cost figure given is for treatment with Debrisan.

2.3. Explicit valuation

Too little costing information was given to comment adequately on this;
it is assumed that the cost for Debrisan was the price charged to the
NHS.

Goode et al. (1979)

3. Allowance for differential timing and uncertainty

Differential timing was not relevant in this study.

Despite the small numbers involved, no sensitivity analysis was carried out.

4. Results and conclusions

The time to secondary closure of the wound was shorter for the group treated with Debrisan, 8.1 days as opposed to 11.6 days. However, the range of observations was also greater, between five and 28 days against six to 22 days for the Eusol group, and the authors noted that the difference in distribution was significant. Three patients in the Eusol group continued to have problems after the wound closure but none of the Debrisan group were affected. The median difference in length of stay was 2.2 days shorter for the Debrisan group.

The cost of the Debrisan treatments was given as £3.40 per day and this was stated to be 'high' but the comparable figure for Eusol treatment was not given. The authors concluded that the high cost of Debrisan was compensated for by savings resulting from shorter lengths of stay but they did not present the actual cost figures to confirm this.

5. General comments

The presentation of this study is rather unsatisfactory as an economic appraisal. Although refinement of costs may not be necessary when relative cost differences are large, the relevant information should be reported even if the figures are crude. In the case of reduced lengths of stay, casual cost saving assumptions are frequently based on average costs, whereas marginal cost savings may be quite small.

The sample used in the study was rather small, particularly in relation to the range of observed outcomes. The authors suggest that a larger study is needed to confirm their results but improved costings are also required. The approach here can be contrasted with a much better appraisal of alternative dressings in Culyer et al. (1983) summarized in this section.

1

Greenfield, S., Komaroff, A. L., Pass, T. M., Anderson, H., and Nessim, S. (1978). Efficiency and cost of primary care by nurses and physician assistants. *New England Journal of Medicine* **298**, 305–9.

1. Study design

1.1. *Study question*

What are the comparative costs of provision of primary care by nurses and physician assistants, and physicians?(a)

1.2. *Alternatives appraised*

Physicians *versus* nurses and physician assistants working with a protocol and physician back-up.

1.3. *Comments*

Four common complaints were studied—respiratory infections, urinary and vaginal infections, headache, and abdominal pain.

2. Assessment of costs and benefits

2.1. *Enumeration*

The costs considered were to the health service (health maintenance organization) only. These included the resources used at the initial visit, including tests requested plus those required for further related visits and hospitalization. Initial costs involved in training staff and setting up the alternative system were not reported.

2.2. *Measurement*

This was a prospective study on 472 patients. About half the patients were randomized (by the receptionist) to either regimen, although scheduling problems prevented total randomization. As the results for the non-randomly assigned patients were not significantly different from the others, these were added to those of the random series. There was no study of long-term health outcome.

2.3. *Explicit valuation*

Market values were used to estimate costs of visits. Practitioner time was of particular interest and this was presented separately.

Greenfield *et al.* (1978)

3. Allowance for differential timing and uncertainty

Differential timing was not relevant in this context. The distribution of each result (mean and standard deviation) was reported and the differences between the two groups were tested for significance.

4. Results and conclusions

The new (nurse, physician assistant) system reduced physician time per patient by 92 per cent, from 11.8 to 0.9 minutes. Average visit costs—including practitioner time and charges for laboratory tests and medications—were 20 per cent less. This was mainly due to lower manpower costs in the new system; the differences in laboratory tests and medications were not significant. When the different complaints were looked at separately, the difference in cost per visit was significant for patients with upper respiratory infections and abdominal pain but not for the other complaints.

5. General comments

This is one of several studies investigating the cost-effectiveness of guidelines or protocols. See also Ruchlin *et al.* (1982) summarized in this section.

2 Gunning–Schepers, L., Leroy, X., and De Wals, P. (1984). Home care as an alternative to hospitalization: a case study in Belgium. *Social Science and Medicine* **18**(6), 531–7.

1. Study design

1.1. *Study question*

What is the impact on health care costs of the establishment of a co-ordinating centre for primary care services?(a)

1.2. *Alternatives appraised*

Establishment of a co-ordinating centre for primary care services *versus* no co-ordinating centre. (See 1.3 below.)

1.3. *Comments*

Apart from the aggregate analysis comparing utilization of services before and after the establishment of the centre, a supplementary analysis of the services received by patients on 'comprehensive home care' was undertaken. These were patients who were adjudged to require hospitalization in the absence of co-ordinated home care services.

2. Assessment of cost and benefits

2.1. *Enumeration*

The costs considered were those resulting from the use of general practitioner services, specialists' services, nursing services, and hospital admissions. In addition, consideration was given to the resources used by those patients receiving 'comprehensive home care'. In considering the costs of caring for patients, a small administrative charge was included to reflect the costs of co-ordination. The resource use of patients and their families was not considered. Also there was no consideration of patient outcome.

2.2. *Measurement*

The analysis was of the 'before and after' type. Data on the utilization of services over a period containing the establishment of the co-ordinating centre were obtained retrospectively from records and analysed for trends. (In addition data on hospital admissions, adjusted by age and sex, were obtained from another location for control purposes.) No statistical tests of significance were reported. Data on the utilization of

Gunning–Schepers *et al.* (1984)

services by patients receiving comprehensive home care were obtained from a diary kept by the head of the home nursing service during one year.

2.3. *Explicit valuation*

Details of the costing procedure for home care services were reported elsewhere. As far as one can tell, these costs were calculated according to the insurance reimbursements made for each service (i.e. market prices). An average cost for hospitalization (per day) was used. The authors point out that this excludes capital building costs, the state interventions to cover the deficits of the hospital sector, and the expenses associated with the technical examinations, which were more frequent in hospital than at home.

3. Allowance for differential timing and uncertainty

Discounting was not relevant in the context of this study since the alternatives have a similar time profile of costs.

No sensitivity analysis was performed. Given the uncertainties surrounding some of the estimates, this was a deficiency in the study.

4. Results and conclusions

The authors suggested that establishment of the co-ordinating centre did not lead to an increase in the use of ambulatory services, such as primary medical care or specialist services. There was an increase in the intensity of nursing services (as measured by the number of consultations per patient) but at the same time a decrease in the hospitalization rate. (This was not equalled elsewhere in the region.) The average cost estimates suggested a difference of Bfr2240 (US$55.17) per day between hospital care and comprehensive home care (for patients who would otherwise need to be hospitalized). However, the authors pointed out that 'the full potential benefit will only be gained if effective measures are taken to reduce the hospital sector'.

5. General comments

This evaluation suffers from a problem common to all of those of the 'before and after' type; that is, can the observed changes be attributed to the intervention in question when other factors have also changed? The authors handle this issue largely in a *qualitative* way, although some figures from Belgium as a whole (on GP consultation rates) and from a similar community (for hospital admissions) are presented. It would have been advantageous to handle some of the uncertainties in a more *quantitative* way, by statistical analysis of trends, or by sensitivity analysis. Particular uncertainties relate to the definition of the reference population (for the medical centre concerned), the potential for consumption of services elsewhere, the resource inputs from the family (in home nursing) and the unreliable nature of hospital average costs.

93 Hallstrom, A., Eisenberg, M. S., and Bergner, L. (1981). Modeling the effectiveness and cost-effectiveness of an emergency service system. *Social Science and Medicine* **15C**, 13–17.

1. Study design

1.1. *Study question*

How cost-effective are emergency medical services (EMS) for cardiac arrest victims?(c)

1.2. *Alternatives appraised*

Basic life support provided by emergency medical technicians (EMTs) *versus* advanced life support provided by paramedics (EMT-para) *versus* rapid defibrillation by EMTs (EMT-defib) *versus* rapid defibrillation by EMTs with paramedic back-up (EMT-defib-para).

1.3. *Comments*

The alternative of doing nothing was not appraised. The first two EMS systems were already operational and provided data for the study. The second two systems were under experimental study and only *expected* effectiveness could be modelled. Regression analysis was used to examine effectiveness and a model was developed which could be used to pre-evaluate the relative effects of programme modifications.

2. Assessment of costs and benefits

2.1. *Enumeration*

Effectiveness was defined as the percentage of patients surviving out-of-hospital cardiac arrest. Cost-effectiveness was taken to be cost divided by effectiveness, expressed as US$1000/% survival. Other outcome effects, such as shorter hospital stay, were not considered. The costs included were for aid units, medical units, overheads and administration of the EMS plus an estimated figure for training the general public in cardio-pulmonary resuscitation (CPR).

2.2. *Measurement*

The authors developed a model in which the probability of survival (effectiveness) was related to the time from collapse to the initiation of CPR and to definitive care. These times were themselves related to features of the EMS and to outside factors, such as the delay in calling assistance and the number of hospitals in the area. Data for the model were obtained from an outcome study of pre-hospital emergency medical care. Patients were not matched but factors such as age and sex had no effect on the probability of survival once the time factors were taken into account.

Cost figures for the existing systems were given in the paper. The cost-effectiveness calculations only considered average cost for each system. The marginal cost per additional percentage survival was not calculated.

2.3. *Explicit valuation*

Details of the sources for cost figures were not given. Presumably, market values were used for cost estimates.

3. Allowance for differential timing and uncertainty

Annuitization of capital equipment was based on a zero discount rate, i.e. straight line depreciation. Two alternative methods of setting up two of the systems were examined, and upper and lower rates of effectiveness and cost-effectiveness were given.

The relationship between turnaround time, effectiveness and cost-effectiveness was shown for one EMS.

4. Results and conclusions

Of the systems already in use, the EMT system was very ineffective, with only six per cent survival to discharge from hospital, compared with the EMT-para system (20 per cent survival). The results from modelling the existing systems and the proposed systems (EMT-defib and EMT-defib-para) gave a range of average costs per percentage surviving from US$52 600 for the EMT system to US$22 200 for the EMT-defib system. The authors concluded that although 'the results for the EMT-defib system were based on supposition, the calculated improvement in effectiveness was impressive and called for a field test of this system'.

5. General comments

Survival to discharge is a limited measure of the success of a service. Other factors, such as post-hospital survival or health status on discharge may be adequately proxied by this one statistic but if this is the case then some evidence should have been given to support this. The impact of the results is reduced by the failure to consider marginal costs. For example, the marginal cost per additional percentage survival for the EMT-defib system compared with EMT alone was US$3000.

94 Lowson, K. V., Drummond, M. F., and Bishop, J. M. (1981). Costing new services: long-term domiciliary oxygen therapy. *Lancet* **i**, 1146–9.

1. Study design

1.1. *Study question*

What is the most cost-effective method of providing chronic bronchitics with long-term oxygen treatment in the home?(c)

1.2. *Alternatives appraised*

Large cylinders *versus* small cylinders *versus* liquid oxygen *versus* oxygen concentrators.

1.3. *Comments*

A distinction was made between costs which were relatively fixed with respect to the number of patients treated and those which were variable. This was important as the fixed costs of one delivery system (concentrators) were higher than the others. (See the comments on marginal analysis in the companion volume, Section 3.2.)

2. Assessment of costs and benefits

2.1. *Enumeration*

The costs considered were those incurred by the health services and by the patients in the provision of oxygen therapy in the home. Other costs and all the benefits of treatment were unaffected by the method provision.

2.2. *Measurement*

A typical pattern of care (15 hours oxygen therapy) was costed as data on individual patients were not available. The effectiveness of oxygen therapy had been assessed in another study.

2.3. *Explicit valuation*

Market values were used to estimate costs.

3. Allowance for differential timing and uncertainty

Most of the costs involved were current costs but the capital component varied between the different delivery systems. The capital costs were converted to annual equivalents by the application of a public sector discount rate at 7 per cent and the rationale for discounting was discussed.

The cost calculations were carried out for different sizes of patient population, to illustrate the effect of fixed costs. Two situations regarding the maintenance of the oxygen concentrators were examined, first, assuming that workshop facilities have to be provided solely for this task and, secondly, that spare capacity exists and only marginal additions have to be made to staff and equipment.

4. Results and conclusions

The cost per annum per patient was given for each method of delivery. For the oxygen concentrators, variation in numbers treated from one to 100 was considered. The figures showed that concentrators were the most cost-effective delivery system for all but the smallest numbers of patients (13 or less depending on maintenance assumptions). The cost per patient per annum was less than £1200 assuming 20 patients, falling to just over £600 assuming 100 patients on the therapy in a given location. The comparative costs were £1486 per patient per annum for liquid oxygen and £2286 and £3640 for large and small cylinders respectively.

5. General comments

The authors discuss other factors which might affect decision making in particular circumstances, for example, the possibility of leasing rather than purchasing equipment. The incidence of costs by viewpoint for the alternative methods is commented on. Although concentrators have been shown to be cost-effective, their provision would fall on a different budget from the costs of the current provision, mainly by cylinders. This creates a barrier to the introduction of a more efficient delivery system. (See the comments on the 'viewpoint for analysis' in Chapter 2.)

95 Ruchlin, H. S., Finkel, M. L., and McCarthy, E. G. (1982). The efficacy of second-opinion consultation programs: a cost-benefit perspective. *Medical Care* **20**(1), 3–20.

1. Study design

1.1. *Study question*

Can second-opinion programmes, by reducing the rate of elective surgery, lead to significant economic savings?(c)

1.2. *Alternatives appraised*

Second-opinion programme (i.e. *mandatory* referral to a second physician to confirm the recommendation for surgery) *versus* no programme.

1.3. *Comments*

The rule was adopted that any recommendation emanating from the consultation process that could potentially lead to reduced health care expenditures was considered to be a nonconfirmation of the original recommendation for surgery, e.g. surgery performed on an ambulatory basis.

2. Assessment of costs and benefits

2.1. *Enumeration*

The costs considered included the administrative costs of the programme, the expenditures for the second-opinion consultation visit and any ancillary tests ordered, plus the costs to participants in obtaining the second-opinion (travel costs, and wages lost due to work time missed). The benefits considered were those resulting from savings in health service resources (surgical and medical care, hospitalization, drug and therapy use, and ancillary service use) and avoidance of work loss and restricted activity days.

2.2. *Measurement*

Data were based on the experience with a mandatory programme administered on behalf of a large Taft–Hartley welfare fund in New York City. During a two-year intake period, 2284 individuals received second-opinion consultations. Of this group, 366 received a nonconfirmation of their needs for surgery. A comparable number of individuals who received a positive confirmation were randomly selected and served as a control for estimating programme savings. Data were drawn from

welfare fund records, supplemented by telephone-administered questionnaires and interviews where necessary.

2.3. *Explicit valuation*

Market prices were used to estimate costs and benefits. These included
physician billings (where these were available—missing items proxied at
US$30 a visit) and hospital *per diem* payments. Data on individuals'
wages were used where available. Where necessary, wages were proxied
based on average wages for the given occupational group. Housewives'
time was valued based on the market valuation of homemaker services.

3. Allowance for differential timing and uncertainty

Costs and benefits were discounted to present values using a 10 per cent
discount rate. A sensitivity analysis was performed using rates of six and
15 per cent.

4. Results and conclusions

Total programme savings were estimated at US$534 791. Of this
amount, medical care utilization savings were US$361 756 and pro-
ductivity savings were US$173 035. The cost of the programme was
US$203 300, yielding a benefit–cost ratio of 2.63. The authors argued
that 'these findings indicate that mandatory second-opinion consultation
programs which are consumer-oriented and intervene before care is
rendered, are clearly cost-effective'. (This procedure is to be contrasted
with the peer review approach, which by design entails a *retrospective*
review.)

5. General comments

See also the editorial comment by Brook, R. H. and Lohr, K. N. (1982).
Second-opinion programs: beyond cost–benefit analyses. *Medical Care*
20(1), 1–2. The major points raised are that (*i*) persons may have a third
consultation to resolve any conflicting messages from the first two, and
(*ii*) evaluations such as the one reported in this paper say nothing about
the impact of second-opinion programmes on health status. For example, does the gatekeeper effect of the programme merely postpone
surgery, perhaps to a less optimal time? This is clearly a subject for
further research. Also it should be noted that actual data were only
available for a proportion of both groups, the remainder being proxied.
However, the authors point out the various assumptions made and the
biases likely to occur.

96 Ruchlin, H. S. and Morris, J. N. (1981). Cost–benefit analysis of an emergency alarm and response system: a case study of long-term care program. *Health Services Research* **16**,1, 65–80.

1. Study design

1.1. *Study question*

What are the costs and benefits of an emergency alarm and response system (Lifeline) installed in the homes of the elderly chronic sick?(d)

1.2. *Alternatives appraised*

Introduction of Lifeline into homes of three target groups *versus* doing nothing.

1.3. *Comments*

A cost–benefit methodology appropriate to the evaluation of long-term care was defined and then applied to the case study of Lifeline. The subdivision of the at-risk population into three target groups allowed some assessment of the marginal impact of extending the scheme to different groups. However, the benefits from averted use of services were valued in average terms rather than marginal.

2. Assessment of costs and benefits

2.1. *Enumeration*

The benefits were defined in terms of averted costs. Three categories of benefits were identified; direct savings from the reduction in use of health facilities and community services; programme externalities, and non-market services, such as voluntary or family support services. Intangibles resulting from enhanced personal security were excluded from the analysis as they could not be quantified. All the direct costs of providing the Lifeline systems were included.

2.2. *Measurement*

An at-risk population was identified and randomly allocated between the control and experimental groups. Matched pairs selected from within these groups were broken down into three target groups, according to social and medical criteria.

The costs of installing and running the Lifeline system were calculated using what was described as 'a task inventory cost methodology' but no details were given. The cost per user was an average cost not marginal.

The authors did not specify how information on service utilization was collected.

2.3. *Explicit valuation*

1977 Massachusetts Medicaid rates were used as the proxies for costs of institutional and formal community support services. Costs per inpatient day and costs per unit of care (e.g. physician unit, meal provision, etc.) were given in the paper. Informal community support services were costed at the prevailing minimum hourly wage rate using Department of Labor data.

3. Allowance for differential timing and uncertainty

No discounting was applied on the ground that the costs only cover one year, yet there were capital costs involved, the benefits of which accrue over a longer period. It was not clear whether capital costs had been annuitized.

The costs and benefits for three target groups, selected according to social and clinical factors, were calculated and the cost–benefit ratios compared.

4. Results and conclusions

The control groups used more days of institutional care than the experimental group, and hence incurred greater costs, US$1 227 000 against US$1 202 000. The control group used more formal community services although the difference was not significant with an associated cost of US$259 000 against US$234 000. The control group also used more units of informal community services but this was only significant for the category of daily checking. The associated cost of informal services was US$32 000 against US$19 000. The experimental group used services costing US$1 455 000 over the 13-month research project, against services costing US$1 518 000 for the control group; the US$62 484 averted cost was attributed to the Lifeline scheme.

The total programme cost was estimated to be US$33 385, giving a benefit–cost ratio of 1.87, over all groups. However, the results were different for the separate target groups. The benefit–cost ratio was less than one for group 1 (severely functionally impaired and socially isolated). For group 2 (severely functionally impaired but not socially isolated) the benefit–cost ratio was 7.19, while for group 3 (socially isolated and either moderately impaired or medically vulnerable) the ratio was 1.27.

Ruchlin and Morris (1981)

5. General comments

The authors begin with a discussion of the relative merits of cost–benefit analysis and cost-effectiveness analysis for evaluating choices in long-term care. However, the definitions used, particularly for cost-effectiveness analysis, are rather different from those which are standard amongst UK health economists.

The use of average costs and the large proportion of averted costs accounted for by institutional care raises the question of whether the benefits suggested are realizable. Care should be used in interpreting the results.

Stilwell, J. A., Young, D., and Cunnington, A. (1980). Evaluation of laboratory tests in hospitals. *Annals of Clinical Biochemistry* **17**, 281–6.

1. Study design

1.1. *Study question*

What are the benefits, in terms of changed patient management, from expenditure on different tests?(c)

1.2. *Alternatives appraised*

The benefits from all test and examination requests were evaluated in the study but only biochemistry tests were costed.

1.3. *Comments*

The study did not have controls but recorded the actions that resulted directly from test information, for a randomly selected group of patients.

2. Assessment of costs and benefits

2.1. *Enumeration*

The costs considered for biochemistry tests were average health service costs, which comprised the costs of collecting blood, processing the samples, and transmitting the results. Included in the costs were technician time, supplies and equipment, overheads, and supervision. The benefits of tests and examinations were defined as the number and type of actions resulting from the information. An action was a change in patient management from what it would have been in the absence of the test result.

2.2. *Measurement*

Costs for biochemistry tests were collected by means of a detailed study of the chemical pathology laboratory in one of the participating hospitals.

Records were kept of tests ordered, the expected results and whether the requests were non-discretionary, diagnostic, monitoring or checking. The results of tests and actions taken were recorded by means of a second questionnaire.

2.3. *Explicit valuation*

Details of the costs used were not given but it is assumed that they were

Stilwell et al. (1980)

wage and price costs to the NHS. Actions were classified as very important, important, trivial or discharge. However, apart from the exclusion of trivial actions, no weights are attached to these in calculating the costs per action for different tests.

3. Allowance for differential timing and uncertainty

The question of differential timing was not relevant.

4. Results and conclusions

15.6 per cent of all tests requested had unexpected results; 13 per cent for non-discretionary tests and 17.8 per cent for discretionary tests. Of the unexpected results, 7.9 per cent led to actions compared with 6.7 per cent of all test results leading to actions. Chest X-ray and ECG had the highest proportion of tests leading to actions, 18 per cent and 11 per cent respectively.

For 42 different biochemistry tests, the expenditure and actions resulting were shown and pounds spent per action (other than trivial) were calculated for the 20 types of test that resulted in actions. The average cost per action for all tests was £22.95. For individual tests where actions resulted, the cost per action ranged from £3.00 to £349.20. Results were presented separately by request category and, in general, the cost per action for tests requested for diagnostic purposes was half that for non-discretionary tests.

The authors concluded that some tests were requested without thought. 192 diagnostic tests had unexpected results but led to no actions. As far as the biochemistry tests were concerned, for a total cost of £1790 'clinicians bought information which uniquely enabled them to discharge five patients, to take 16 courses of action which would have had very serious consequences if omitted, and seven actions which would have had serious consequences'.

5. General comments

Strathclyde Diabetic Group (1983). Disposable or non-disposable syringes and needles for diabetics? *British Medical Journal* **286**, 369–70.

1. Study design

1.1. *Study question*

Is the use of disposable syringes and needles by diabetics cost-effective?(c)

1.2. *Alternatives appraised*

Disposables (single use) *versus* disposables (with re-use) *versus* non-disposables.

1.3. *Comments*

It was thought that the re-use of disposable syringes and needles was the cheapest option. The main purpose of the study was to establish the safety (i.e. effectiveness) of re-use.

2. Assessment of costs and benefits

2.1. *Enumeration*

The only costs considered were those of the disposable or non-disposable equipment and the supplies required for sterilization. Patients' time and convenience were not considered. Effectiveness was assessed by measuring the incidence of infections.

2.2. *Measurement*

Information was collected by means of a questionnaire to adult diabetics. They were asked about the equipment they used, their pattern of re-use of disposables or sterilization practices, and whether they had experienced infection at the site of an injection. This information was obtained retrospectively and was not independently confirmed. The patients do not appear to have been asked about the frequency of any infection.

2.3. *Explicit valuation*

Equipment and supplies were valued at hospital contract prices exclusive of value added tax.

Strathclyde Diabetic Group (1983)

3. Allowance for differential timing and uncertainty

Discounting was not relevant.

The authors reported the distribution of responses to the questions on use of equipment, although the costs were only calculated for the mean result.

4. Results and conclusions

Of the patients who were already using disposables, 51 per cent re-used their needles and 65 per cent re-used their syringes. The mean annual use was 184 syringes and 234 needles. If patients who discarded their disposables after one or two days because of the instructions to 'use once only' were excluded the mean figures fell to 109 syringes and 122 needles. The comparable annual figures for non-disposables were 1.72 syringes and 62 needles.

The presentation of the cost figures was rather confusing, as there was no explicit comparison for these actual patterns of re-use. Single use of disposables instead of non-disposables would result in additional costs of either £13.94 or £22.22 per patient per year depending on the type of disposables used. The cost for reuse of disposables was based on 'best practice' requiring 26 syringes and 120 needles per year. The annual saving would be between £4.92 and £9.17 per patient. The average practice of 184 syringes and 234 needles would result in higher equipment costs than non-disposables but this figure was not reported. The authors' conclusion was that 'any re-use of disposable equipment is equal to or cheaper than the present non-disposable equipment'. However, they do not indicate how they would ensure patient compliance with re-use of disposables.

On the effectiveness of re-use, there was one infection reported amongst 153 patients re-using disposable equipment, whereas 490 patients sterilizing non-disposables according to approved methods had seven infections. This appeared to indicate that re-use was equally effective, but it should be noted that the re-use group was self-selected and might be more careful about avoiding infections than a randomly selected group. 76 per cent of patients expressed a preference for disposables even if they had to be re-used.

5. General comments

The results should be treated as indicative of possible improvements through the use of disposables but the study would be more robust if a prospective survey had been conducted with random selection of patients. However, the costs of non-disposables are underestimated because patient time has not been included. This is likely to be one factor in the majority preference for disposables.

99 Woodward, E. G. and Drummond, M. F. (1984). Cost-effectiveness in a hospital contact lens department. *Ophthalmic and Physiological Optics* **4**(2), 161–7.

1. Study design

1.1. *Study question*

How can the cost-effectiveness of providing contact lenses to ophthalmology patients be improved?(c)

1.2. *Alternatives appraised*

Purchasing more lenses externally *versus* internal production. Holding stocks of lenses *versus* making lenses to order.

1.3. *Comments*

The authors addressed two specific questions as examples of the issues relating to the cost-effectiveness of contact lens provision. They do not claim that these are comprehensive; rather the opposite. The study is careful to identify and measure the appropriate marginal costs.

2. Assessment of costs and benefits

2.1. *Enumeration*

The costs involved in the production of lenses were for staff, equipment, materials, and laboratory space. For the purchase of lenses, the only cost considered was the purchase price; estimation of production cost was not attempted. Holding a stock of lenses involves costs for administering the system. Against this is set the savings to the health service and to the patient from avoided second visits.

2.2. *Measurement*

The costs and savings associated with stock holding vary with the level of stock held. The authors were careful to identify marginal costs and benefits associated with different levels of stock held. Estimates of numbers of patients who could be fitted from stock were based on workload figures for the previous six months. Savings to patients were based on a survey of 100 patients.

2.3. *Explicit valuation*

Market prices were used where they were available; NHS wages and prices, purchase cost of lenses, out of pocket expenditure by patients, and wages lost. The rental value of building space was imputed. The cost of patients' time (apart from lost wages) was not discussed.

3. Allowance for differential timing and uncertainty

Interest charges were assessed on capital used in internal production. It is not clear why this particular procedure was used, rather than simple annuitization.

Given the nature of the results, sensitivity analysis was not necessary.

4. Results and conclusions

Internal production costs were lower than the cost of buying in for all forms of lenses, although the cost differences varied. This result held even when imputed rents and interest charges were added to the cost of internal production.

The holding of stocks of lenses was found to be worthwhile for one group of patients (aphakic) whose lens needs were sufficiently similar. Even so, an initial system of 35 different lens categories was suggested, allowing an estimated 75 per cent of these patients to be fitted from stock. At this level, the system could be administered by redeploying staff time. In addition, there would be a saving of 200 clinical sessions. Patients would save an average of 3.87 hours plus out-of-pocket expenditures and lost wages. In addition, these patients would benefit by having their lenses fitted four to six weeks earlier.

5. General comments

The authors discuss the implications of various developments in lens production and use and identify other aspects of ophthalmic care that might be amenable to economic appraisal in the future.

100 Zapka, J. and Averill, B. W. (1979). Self-care for colds: a cost-effective alternative to upper respiratory infection management. *American Journal of Public Health* **69**(8), 814–16.

1. Study design

1.1. *Study question*

Is it cost-effective to establish a cold self-care centre (CSC) to handle patients with upper respiratory infections in an ambulatory care setting?(c)

1.2. *Alternatives appraised*

CSC (using algorithm for self-assessment with nurse practitioner support) *versus* normal outpatient care.

1.3. *Comments*

2. Assessment of costs and benefits

2.1. *Enumeration*

The costs considered were the resources required to develop the CSC and the resources used in operating it, and the resources used in operating normal outpatient care. The benefits considered were provider and patient satisfaction and patient knowledge and behaviour.

2.2. *Measurement*

This was a 'before and after' study. No prospective controlled evaluation was carried out. The reported visits for the relevant illnesses were monitored for three years before and two years after in order to assess the impact of the CSC. No assessment of health outcome was attempted, other than indirectly by reference to providers' views.

2.3. *Explicit valuation*

Market values were used to estimate costs and savings from reduced visits. No attempt was made to value benefits.

3. Allowance for differential timing and uncertainty

Not considered. Differential timing would be relevant as the savings, from reduced visits, occur in the future. Also, given that there were a number of ways of using resources freed by the reduction in visits, a sensitivity analysis of the difference options would be of benefit.

4. Results and conclusions

There were no adverse comments from providers or patients in respect of the new scheme. Use of the CSC did not appear to change attitudes or self-medication behaviour, but it affected care-seeking behaviour. 20 per cent of users referred themselves immediately to professional care and 6 per cent anticipated seeking professional care for any subsequent cold. Over the two years reductions in visits to clinicians for the conditions concerned, if attributed to the CSC, meant that about US$46 000 was saved by its introduction. (See 5 below.)

5. General comments

It should not be assumed that the US$46 000 savings were generated in money terms. It may be that clinicians' time freed from these visits was allocated to other activities.

As patients were not randomly assigned to the CSC this study only investigates the impact for those patients who were predisposed to use it. These demonstrated higher knowledge of cold care than non-users. The approach adopted has relevance in planning terms, however.

Indexes

In the indexes, reference is made to the study number not page number. As well as being indexed by author, studies are grouped below in a Medical Index. Therefore it is possible to locate all the studies in the field of (say) cardiology by consulting the Medical Index. A Phase of Treatment Index was also included in the first volume of Studies. Obviously, this is not necessary in this volume, since the phase of treatment (e.g. diagnosis) is self-evident from the sectioning of the book.

We are grateful to Dr Dorothy Kyle and Miss Sue Elias for help in compiling the indexes.

Medical index

Author index

302

303